Environmental Policy and Industrial Innovation

Research by the Energy and Environmental Programme is supported by generous contributions of finance and professional advice from the following organizations:

Amerada Hess • Arthur D Little • Ashland Oil
British Coal • British Nuclear Fuels
British Petroleum • European Commission
Department of Trade and Industry • Eastern Group
Enterprise Oil • ENRON Europe • Exxon • LASMO
Mobil • National Grid • National Power • Nuclear Electric
Overseas Development Administration • PowerGen
Saudi Aramco • Shell • Statoil • St Clements Services
Texaco • Total • Tokyo Electric Power Company

Environmental Policy and Industrial Innovation

Strategies in Europe, the USA and Japan

David Wallace

Earthscan Publications Ltd, London

First published in Great Britain in 1995 by
Royal Institute of International Affairs, 10 St James's Square, London SW1Y 4LE
(Charity Registration No. 208 223) and Earthscan Publications Ltd,
120 Pentonville Road, London N1 9JN

Distributed in North America by
The Brookings Institution, 1775 Massachusetts Avenue NW,
Washington DC 20036-2188

© Royal Institute of International Affairs, 1995

All rights reserved.

A catalogue record for this book is available from the British Library.

Paperback: ISBN 1 85383 288 X
Hardback: ISBN 1 85383 289 8

The Royal Institute of International Affairs is an independent body which promotes the rigorous study of international questions and does not express opinions of its own. The opinions expressed in this publication are the responsibility of the author.

Earthscan Publications Limited is an editorially independent subsidiary of Kogan Page Limited and publishes in association with the International Institute of Environment and Development and the World Wide Fund for Nature.

Printed in England by Clays Ltd, St Ives plc
Cover by Visible Edge.
Cover illustration by Andy Lovel.

To Anita

I am sorry to have put
you through this.
It's your turn now.

Contents

Part 4

Tables

Figures

Boxes

Foreword

Whether environmental policy hurts or helps international competitiveness and long-run economic development is a topic of considerable debate. Few parties to that debate have been clear in recognizing the central role that innovation plays in determining the answer, and the extent to which the effects of innovation are inherently unpredictable. In this study, David Wallace argues that in view of this unpredictability, and wide variations in national conditions and requirements, the impact of innovation on competitiveness is in general indeterminate and he focuses instead on a far more productive question: what can the experience of environmental policy in the OECD tell us about the types of policies and strategies that maximize innovation in pursuit of environmental goals?

In pursuing this question, Wallace here brings together exceptional insight into the different policy and wider cultural assumptions that underpin the relationship between government and industry in the six countries studied, particularly – but not exclusively – as it affects the quality and consistency of environmental policy. And, in his case studies of the development of 'clean cars' and emission controls for power stations, he has produced unparalleled analysis of just how environmental policy in the different countries did – and did not – contribute towards more environmentally effective and economically competitive responses.

This study is the third with a specific 'industry and environment' theme published by our Programme in recent years. Its strategic insight into the nature of the relationship between regulator and industry, and its systematic exploration of the differences between OECD countries, help to provide a framework for understanding the industrial perspectives surveyed in *Environmental Profiles of European Industry*, and for assessing the different industrial and policy positions set out in *Sustainable Development and the Energy Industries*.

To carry out such an ambitious project, we are fortunate indeed to have had a researcher as capable as David Wallace. He has worked hard to bring some order and comprehension to a seemingly chaotic set of national and industrial experiences, and readers will find a refreshing flair for making sharp and sometimes acerbic judgments about the prejudices, successes and failures that have permeated national experiences of environmental policy. We are indebted to the UK Cabinet Office for granting the fellowship that has supported David in his studies, and to the DTI for giving David the time, scope, support and intellectual freedom required. But our biggest thanks, of course, must go to David himself for pursuing his interests and ideas to fruition.

June 1995 Michael Grubb
Head, Energy and Environmental Programme

Acknowledgements

This book is the outcome of a fellowship extended to me by the UK Cabinet Office. My gratitude goes firstly, therefore, to all those in the Cabinet Office who devised and organized the Government Research Fellowship scheme. The Department of Trade and Industry has provided me with great support from the outset and has continued to show generosity and patience by allowing me time to complete this book. I hope they will feel I have used this opportunity to step back from the day-to-day concerns of Whitehall wisely. If in the process I have sometimes gone too far and stepped off the edge of a cliff, it is entirely my own doing.

Michael Grubb took me in as a visiting researcher at the Royal Institute of International Affairs. There can be no better place to conduct a study such as this. The advice and assistance of Michael, Walter Patterson, Stephen Peake, Matthew Tickle, Jane Chapman, Nicola Steen and Nicole Dando have been invaluable. Involvement in workshops and meetings and contact with other visiting fellows, especially Yukinoro Nakano and Hirotaka Ono, helped to broaden my understanding of previously unfamiliar areas. During the course of two study groups many individuals contributed ideas and advice. I am also indebted to Josh Epstein, Rob Axtell and John Steinbruner at the Brookings Institution in Washington D.C., where I spent several months as a guest, enabling me to tackle the complexities of the United States at enjoyable leisure.

Ultimately, this book is founded on the views and material freely given by all those interviewed for the national and industry case studies. Too numerous to mention, they were without exception accommodating and eager to help my understanding of their motivations, policies and practices.

May 1995 David Wallace

About the Author

David Wallace joined the Department of Trade and Industry in 1986 as an oceanographer developing models of oil and chemical spills. From 1990 to 1993 he worked in the Department of Energy (latterly subsumed into the DTI) with responsibilities for international relations in energy technology. In 1993 he took up a position as a visiting Research Fellow at the Energy and Environmental Programme of the Royal Institute of International Affairs, following the award of a Government Research Fellowship by the Cabinet Office.

Summary and Conclusions

Innovation generally lowers the cost of responding to a change in the commercial environment. Environmental policies have greatly altered the commercial conditions for most industries over the past few decades. In some countries, policy-making processes have given industry both the freedom and the incentives to make innovative responses to environmental challenges. In others, policy-makers have failed to recognize either the necessity or the conditions required for innovation; political circumstances have pitted industry against legislators and regulators, regulatory mechanisms have been inflexible and industry's costs for cleaning up the environment have been higher than necessary.

The policy and regulatory regimes which are most successful in harnessing innovation have several features in common. The essential characteristic is to establish a high quality, honest dialogue that nevertheless does not compromise the independence of environmental policy-making from industry's special interests.

When policy-making has strong political independence from industry (e.g. due to the influence of environmental pressure groups), but dialogue is poor, environmental regulations are associated with high compliance costs. This can create a vicious circle, in which political polarization of environmental issues leads to less dialogue, resulting in poor and costly regulations and hence further polarization. However, political independence combined with good dialogue allows policy-makers to develop flexible regulatory mechanisms and schedules which accommodate innovation.

Conversely a close relationship that compromises the political independence of policy-makers will mean that environmental costs are not adequately reflected in decision-making. Industry has the means to stifle or delay environmental policy and the potential for innovation may not be fully considered. Yet this close political association can be accompanied by good dialogue, leading to regulatory mechanisms which allow innovation to play a role. Those environmental goals which do gain political backing can be pursued at least cost to industry.

National case studies reveal how this complex of relationships has developed in different OECD countries. In the United States, an adversarial approach to public policy has exacerbated the tensions which accompany environmental issues. Political toughness has found expression through poor relations between policy-makers and industry, leading to an inflexible and costly regulatory structure. Germany reveals similar difficulties, a situation which is at odds with generally good state–industry relations there, on issues other than the environment.

Japan has developed a strong sense that environmental goals should not be compromised by narrow industrial interests, in large part through the efforts of powerful local authorities. A web of communication, backed by formal processes of information exchange, allows for flexible, adaptive environmental goals which progress from indicative, trial targets, through to local and then national ordinances. Denmark has taken a similar tough stance, bolstered by general assent from industry which has the comfort of being closely integrated into policy-making processes. In the Netherlands, two recent mutually supportive policy initiatives – a comprehensive, long-term, national plan for sustainable development and widespread use of contracts ('voluntary' agreements) between firms and government – may lead quickly to the kind of constructive relationships which have evolved more gradually in Japan and Denmark.

In France, a technocratic elite ensures that industry retains some of the special political significance attached to it in the postwar period of rapid growth. Environmental issues struggle for political credence, by comparison with industrial interests. However, the technocratic approach to policy-making tends to produce a regulatory approach in tune with industry's capabilities and capacity for change and innovation.

Two industrial case studies shed light on the ways in which politicians and industrialists have made connections between environmental issues and fears of losing competitive advantage to other countries. In both cases the sources of innovation and its impact on industrial competitiveness have been misunderstood, and international trade and environmental issues have been needlessly politicized.

Japanese vehicle emission standards were resisted by industry and government in the United States in the late 1970s. The Americans feared the costs of adapting to new catalytic converters and suspected the Japanese of subsidizing their growing exports to the United States. In fact, Japanese manufacturers had developed the system of lean production. Superior to mass production, it allowed them to drive down costs, rewarded continuous innovation and simplified production changes.

In Europe a similar story unfolded in the 1980s. Politicians thought that new requirements for catalytic converters, rather than lean-burn engines, would favour German car manufacturers and seriously disrupt, or even finish off, the ailing British car firm, Rover. Fortunately, Rover made the switch to lean production with the help of its Japanese partner, Honda, and then took catalytic converters in its stride and prospered while German firms suffered badly in Europe's recession.

Poor dialogue between industry and government was partly to blame for the American attitude towards the Japanese emission standards. More recently, Californian policy-makers have been misled by, and have misunderstood, the US vehicle manufacturers. As an unintended consequence, the new regulation requiring electric vehicles to be supplied in large numbers by 1998 is dependent on basic technological breakthroughs which may never occur. This exception proves the rule that policy-makers generally try to force not technology but the commercialization of technology.

Acid rain from power plants can be reduced with expensive desulphurization and NOx removal technology. These technologies are big business for large, politically influential engineering firms. Hopes – and fears – that international agreements on acid rain would give a dominant market position to engineering firms from countries which have previously taken unilateral action to reduce acid emissions (such as Germany) have proved unfounded. Fuel switching and closure of coal power plants has reduced emissions with little need for control technology to be retrofitted to existing plants. In addition, the history of the development of control technologies illustrates that they need to adapt to local market conditions. Local adaptation has allowed domestic firms to take over the leading edge of the technology, even where it has originally been licensed from foreign competitors. It is likely that in the future any large new market, e.g. China, will inspire such local adaptation.

So, do tough environmental standards at home lead to competitive advantage abroad, as European and American pundits and politicians have suggested? The acid rain experience suggests that international environmental negotiations, market structures and technological trajectories are unpredictable: a national 'first-mover' strategy is not a sensible basis for environmental policy-making.

Recently, policy-makers have become interested in the merits of economic instruments vs. standards-based regulations and some have been led to believe that taxes and pollution charges invariably provide superior incentives for innovation. This conclusion is unsupported, in theory or practice, and would on its own make a poor basis for policy. Instead, decisions should be based on international experience to date with environmental policy: firms are more comfortable innovating when risks are reduced and risks are lower when environmental policy is stable and credible over the long term, and when regulatory processes are based on open, informed dialogue and executed by competent, knowledgeable regulators.

The long term challenge of sustainable development is an opportunity for governments to make environmental policy more stable and less reactive. New industry–government working relationships such as flexible 'voluntary' agreements and contracts are devolving greater responsibility to firms, while increasing dialogue. This leads to more flexibility to innovate, lower compliance costs and less opposition to environmental policies. Sustainable development will require such politically sustainable environmental policies and continuous pressure, and opportunity, for industry to innovate.

Part 1

Chapter 1

Introduction

1.1. Purpose of the Study

Innovation is the one constant factor in wealth creation. When populations stabilize, when consumption of raw materials and of energy levels off and when there are no new lands to discover, our ability to find new, useful activities and products and to improve upon existing ones continues to enrich our quality of life. Those individuals, corporations and nations which show the greatest capacity for innovation, and exploit that capacity to the best effect, will be rewarded by the rest of us for their efforts.

This study explores the links between environmental policy and industrial innovation. It seeks out lessons for policy-makers, for industrialists and for all who wish to understand how to reduce our impacts on the environment with the least possible restriction on economic growth and our quality of life. The purpose is not to produce a universally applicable list of the types of regulatory tools available to policy-makers, ranked in order of their impact on innovation. I believe that such a list would not really be credible. Instead, the real lessons for policy-makers are to be found in the histories of environmental and industrial policy in the developed countries and especially in the political, institutional and personal relationships in which these policies are embedded and through which they are applied. They are lessons which can be applied in countries which have in the past failed to understand the needs and role of innovation, and in developing countries which are beginning to face up to the tensions between environment and development.

1.2. Policy Background

At the national level, governments make decisions which directly, and more often indirectly, influence the range and effectiveness of innovative activities within their domestic firms. This perception has grown since at least the late

1970s, when deregulation became a political issue in Britain and the United States and policy-makers began to take a keen interest in the effects of regulations on industry. One of the major areas of government regulation of industry is environmental protection and, among the possible effects on industry, one of the major areas of concern is the impact on innovation.

Deregulation became a pressing issue during the Reagan and Thatcher era because industry was telling government that regulations were imposing excessive costs on firms. In passing – and with no shortage of anecdotal evidence, especially from the United States – policy-makers came to assume that innovation was also suffering from these regulations. It was but a short step to the conclusion that regulations stifle innovation. Even so, the financial burden, i.e. cost, of complying with regulations was the main area of concern. Higher costs mean dearer products, dearer products mean fewer sales. Surely, therefore, at a national level, competitiveness must suffer by comparison with other countries, trade deficits will grow and politicians will panic?[1]

International competitiveness has become the holy grail of politicians in the developed world. In the 1980s America was convinced that it was losing competitiveness against Japan. Europe, in turn, agonized over its lack of competitiveness against Japan and the United States. Britain began the 1980s believing it was losing its competitiveness against every other country and ended the decade with many briefly believing that its situation was miraculously transformed. Throughout this period, the dynamic Asian economies such as South Korea and Taiwan flourished, sowing seeds of doubt about the long term competitiveness of all the OECD countries.

At first, concern about industrial innovation was taken along for the ride, as it were, by the far greater concern over costs and competitiveness. But gradually politicians and policy-makers, especially in the UK, began to realize the fundamental importance of innovation for long-term, sustained competitiveness. Increasingly, the effects of government policy on innovation became interesting in their own right.

---

[1] In fact, most studies consistently fail to find any significant link between environmental regulations and trade. For a good recent study see Robert Repetto, *Jobs, Competitiveness and Environmental Regulations: What are the Real Issues?,* World Resources Institute, Washington DC, 1995.

1.3. The Revisionist View of Environmental Regulations

Out of these simultaneous preoccupations with competitiveness, innovation and regulatory burdens, there emerged, around 1990, the 'revisionist' view of environmental regulations. Put simply, this asserts that environmental regulations can be good for a country's competitiveness.[2] A major plank of the revisionist view is the assertion that tough regulations will stimulate innovation, making firms generally fitter and more competitive. According to revisionists such as Michael Porter of the Harvard Business School, tough environmental standards can lead to a national competitive advantage in two ways.[3]

1. The First-Mover Strategy

A country may set tighter environmental standards than elsewhere, forcing its industry to develop improved processes or dedicated pollution control equipment. If other countries subsequently adopt similar standards, and trade is unrestricted, the firms in the country which first applied the standards are likely to dominate the market for the associated technologies. The apparent potential of this first-mover strategy has been the basis of the Clinton administration's public statements of support for high environmental standards in the United States, and similar statements by politicians elsewhere.

2. Stimulating Innovation

The proposition here is that tough environmental standards stimulate industrial innovation. This results in superior technologies and improved corporate performance, bringing competitiveness benefits which outweigh any additional costs of meeting the increased standards.

In support of the revisionist view, Porter states, rather boldly, that the existence of tough environmental regulations in Japan and Germany, where GNP growth and productivity growth had recently been higher than in the United States, is 'the strongest proof that environmental protection does not

---

[2] The label of 'revisionist' was coined in Karen L. Palmer and R.D. Simpson, 'Environmental Policy as Industrial Policy', *Resources*, Summer 1993.

[3] See, for example, Michael E. Porter, 'America's Green Strategy', *Scientific American*, April 1991, p. 96.

harm competitiveness'. However, he draws a distinction between good regulations – which use market incentives, are sensitive to costs and stress pollution prevention – and poor regulations, which constrain the choice of technologies and stress end-of-pipe and clean-up measures. American regulations are held to have been of the poor variety.

The revisionist view of environmental regulations has had considerable influence. Al Gore expressed revisionist ideas frequently during the US presidential campaign which saw him installed as Vice-President to Bill Clinton in 1992. The Business Council on Sustainable Development reflected these arguments in its milestone publication Changing Course.[4] According to some of those interviewed for this study, many of the US Environmental Protection Agency's initiatives since 1992 are inspired by a belief that appropriate regulations can improve the competitiveness of US industries. In France, Edith Cresson put great emphasis on the industry-strategic use of environmental regulations, when Minister for Foreign Trade and Prime Minister.

1.4. Limitations of the Revisionist View

The appearance of the revisionist view has played a part in persuading policy-makers to think again about the effects of environmental policy. More importantly it has brought greater prominence to the question of innovation and environmental policy (and inspired this study). Unfortunately, the question implicit in the revisionist view – Which types of regulation are good for innovation? – is too limited to be of much use to most policy-makers and industrialists.

Innovation is a complex process, encompassing everything from basic research activity, to new working practices or even more attractive packaging. Chapter 2 outlines a perspective in which innovation is inseparable from the organizational and managerial structures and capabilities of the firm. This view stresses that innovation is influenced by the firm's perceptions of the business environment and its corporate strategy. An analysis which attempted only to map details of regulation (what type of regulatory instrument?; applied when?; emission levels?; tax rates?) to responses within the regulated industry

[4] Stephan Schmidheiny, *Changing Course: A Global Business Perspective on Development and the Environment*, MIT Press, Cambridge, MA, 1992.

would fail to capture many other relevant aspects of environmental policy. A limited analysis of this kind might succeed in generating robust quantitative answers about R&D expenditures or tonnage of patents but these are poor proxies for innovation (for example, in those firms which are innovative but have no formal R&D programme).

What is required is an approach which encompasses the complexity both of innovation and of the political context, formulation and execution of environmental policy.

This study does not examine the other element of the revisionist view, the first-mover strategy, in any great depth. That strategy is not directly related to innovation, being more concerned with market analysis and the timing of product development. However, real or imagined first-mover strategies were part of the policy backdrop for some of the examples used in this book, including the case study of coal power stations. In most cases, the first-mover strategy (if it ever truly existed) has been blown off course by other political, economic or commercial developments. Policy-makers seem to have an instinctive feel for the fragility of the first-mover strategy, as none of those contacted during the course of the study would give it their backing.

1.5. Harnessing Innovation

The history of economic development tells us that innovation is the key to sustained wealth creation. It is intuitively obvious that in any endeavour we should try to ensure that innovation is allowed to play a role. In pursuing environmental objectives we should allow innovation to play as large a part as possible, by providing the opportunity and incentives for firms to innovate.

Chapter 2 indicates how the perceptions, behaviour and organization of firms interact with their commercial environment and determine the conditions for successful innovation. This provides the link with the actions of policy-makers, who influence the political background of environmental issues and the methods for formulating and implementing environmental goals. Studies of different national approaches to environmental policy (Chapters 3–8) and industrial case studies (Chapters 9 and 10) lead to the conclusion that the political, institutional and even personal relationships between policy-makers and industry have the greatest bearing on innovation.

1.6. Case Studies

1.6.1. Country Case Studies

Chapters 3–8 describe those aspects of environmental and industrial policy and industry–regulator relationships which influence industrial innovation in six countries: Denmark, the Netherlands, Germany, France, Japan and the United States. The emphasis is unashamedly on the developed Western economies, where substantial government interest in environmental policy can be traced back over at least two decades.

The choice of Europe, the United States and Japan covers the 'triad' of regions which is commonly used to distinguish the world's three major trading blocs. Four European countries were studied in order to reflect some of the important variations in policies and attitudes within the region. Germany was a natural choice as Europe's biggest and most successful industrial nation. Of the other major European nations, France demonstrates a very close, sometimes inseparable relationship between state and industry and so can be expected to provide a stark contrast with the United States in particular. Further choices were influenced by the likelihood of encountering well focused environmental and industrial policies, and policy-makers and industrialists who could be expected to elucidate coherent views. Denmark was chosen as just such a place, with the additional advantage that, as a relatively small country, it might provide a particularly simple study with which to introduce some of the salient issues. The Netherlands was the final choice due to its reputation for pursuing radical, innovative industry–environment policies in recent years.

The European study visits typically lasted three to five days and consisted of face-to-face interviews with policy-makers and industrialists, often represented by major industrial associations. Government contacts were mainly located within the core ministries of interest – environment, industry and on occasion energy and research. Typically, these were individuals with responsibility for technology within environment ministries and responsibility for environmental issues within industry ministries. They were commonly heads of major divisions within ministries (under secretary) or one rank below this (assistant secretary). Only rarely did individuals have specific responsibility for considering the impacts of regulations on innovation, and while some interviewees were unwilling to depart far from descriptions of existing policies and procedures, most were happy to discuss their views of issues such as the relationships between ministries, industry and the public, and to offer their appraisal of the successes and failures of past, present and future policies.

In the United States, the study of national policy was supplemented by a study of the situation in California. That state was chosen because it has extreme environmental problems and because regulators there have claimed that some of their regulations have been intentionally 'technology-forcing'. More time was available for the United States study than for the other country studies. This proved to be of immense value as it allowed for the special focus on California.

1.6.2. Industry Case Studies

The country case studies are followed by two industrial case studies, of passenger vehicle emission controls (Chapter 9) and coal power station emission controls (Chapter 10). The environmental problems which forced politicians to act were, in both cases, the defining environmental issues of their time. For vehicles in the 1970s the issue was urban air quality. For coal power generation some years later the issue was acid rain. In addition, both of these industries are global, with significant presences in Europe, Japan and the United States, so they cut horizontally across the individual country studies and provide an opportunity both to look deeper into national approaches and to make international comparisons.

Vehicle emission regulations have been a matter of contention between the United States and Japan since the early 1970s and became a significant political issue within Europe in the 1980s. Accusations of trade motives behind new regulations and fears of losing competitiveness were rife, particularly in the United States. Studying this industry provides insights into the proposition that manufacturers become more innovative when challenged by regulations and of the converse, orthodox view that regulations are simply a burden which impose increased costs.

The world market for power station emission control equipment is very large, and as such seems a perfect candidate for governments to engage in the first-mover strategy. As with vehicles, governments have accused one

Methodology used for the interviews

Individuals were identified in advance but initial interviews often led to further recommendations of contacts which were pursued wherever possible. The assistance of the staff of the UK's embassies in each of the countries I visited was invaluable throughout this process. The interviews lasted between 30 minutes and several hours. Contacts were sent a list of issues in advance but these in no way constrained the interviews. The variations between national attitudes and approach were so great that I was obliged to let the interviewees set the agenda to some extent, in order to reflect what they felt were the important issues. On occasion, I have found it useful to quote from some of these interviews but individuals are not identified, unless they fill a public, political role, and were speaking in that capacity (e.g. White House officials in the United States). However, quotes and opinions are generally accompanied by an indication that they come from a government official and sometimes I have found it useful to indicate the particular ministry.

Material was sent in advance to the interviewees, alerting them to the issues to be discussed. The following issues were common to most of the interviews:

- structure of environmental policy-making; nature and timing of consultation with industry;
- nature of relationships between arms of government and industry;
- examples of policies/regulations being influenced by trade or competitiveness issues;
- interviewees' understanding of impacts of regulations on technological development; examples of negative/positive impacts.

Although the interviews were the key to understanding government–industry relationships, the case studies also draw on many published sources of information. Government publications and policy studies from academia, research institutes and pressure groups have contributed to the background descriptions of national environmental and industrial policies. These sources are referenced in footnotes throughout the text.

another of having ulterior motives behind their pursuit of environmental regulations, particularly international regulations. This was most notably the case in Europe during the 1980s, when the main protagonists were Germany and the United Kingdom.

The methodology used in the industry studies is similar to the approach taken in the country studies. Individuals with responsibility for environmental regulations and industry policy relating to vehicles and power stations were identified and interviewed during the study visits to each of the six countries, where this was relevant. (Denmark, for instance, has no significant vehicle industry.) Inevitably, some interviews touched on both the industry and country studies: interviewees would discuss specific industry topics and also deal with broader national policies and attitudes, often referring to vehicles or power stations by way of illustration.

1.7. Synthesis and Analysis

The focus of this study was dictated by a strong sense that the degree to which policy-makers are cognisant of, and sensitive to, the needs of industry varies enormously and that there are valuable lessons to be learned from different national approaches. This perspective led to the following issue being addressed – how can innovation best be harnessed to meet environmental objectives?

The analysis following the country and case studies makes recommendations for the strategic management of the environment–industry policy interface, with the aim of harnessing innovation to the task of achieving environmental objectives. This final section identifies the approaches of the six countries with several distinct national-level strategies, some of which will be more attractive than others to rational policy-makers. The analysis identifies certain features of the industrial and environmental policies encountered in the case studies which appear to increase the chances of pursuing particularly successful strategies.

In the course of this study it became clear that, in certain countries, national efforts to implement sustainable development are revolutionizing industrial and environmental policy. If these policies are maintained, these economies are likely to be subjected to extraordinary demands which can only be met by successfully harnessing innovation. The lessons and analysis in this book may go some way towards meeting this need.

Chapter 2

Policy and the Innovation Process

2.1. Introduction

In order to explore the links between environmental policy and industrial innovation, we must first look at innovation itself. Definitions of innovation abound, and throughout this book a very broad one will be used. Innovation is approached from the perspective of the firm rather than the individual researcher in a government or industry laboratory. From this viewpoint 'innovation' can encompass any change in technology, production processes or organizational and managerial structure and techniques.

This firm-centred, rather than research-centred, view of innovation has been chosen for a good reason. The purpose of this study is not to quantify research and development expenditures but to identify those features of environmental policy-making and regulation which appear to influence the opportunity, willingness and capacity of firms to choose innovation as part of their response to environmental goals.

This chapter explores the firm-centred view of innovation. It reviews relevant literature, mostly from organizational theory, on the internal, operational features of firms which are successful innovators. Environmental policy creates markets and influences the commercial environment of the firm. The consequences of this for a firm's perceptions of risk, and for its strategic decision-making (and particularly the decision to innovate) are drawn out. Towards the end of the chapter, these factors are related to aspects of policy where governments can be expected to exercise some control, particularly as regards industry's involvement in regulatory processes and the overall political climate of environmental policy.

This chapter also examines the limitations of economic theory in this area. Since the 1970s, economic theory has been a major source of policy advice on the choice of least cost and socially optimal regulatory instruments, and later, through familiarity, on the choice of instruments best suited to encouraging innovation. However, economists are now overturning earlier results, leading some to conclude that environmental economics is not a reliable source of policy advice when innovation is considered to be important, i.e. in all but the most short-run policy decisions.

2.2. Processes of Innovation in the Firm

For many policy-makers, innovation is still synonymous with research and development carried out in laboratories by boffins in white coats. This view persists although the linear technology-push' model underpinning it was abandoned by researchers into innovation during the 1970s. Equally outdated is the (equally linear) 'market-pull' model of innovation, which imagined innovation to be simply a matter of allocating research and development resources to areas identified by market research. The development of models of industrial innovation has been summarized by a long-standing investigator in this field, Roy Rothwell of the UK-based Science Policy Research Unit.[1]

Rothwell highlighted the internal factors for successful innovation commonly identified by systematic studies undertaken since the 1950s. He found that successful innovative projects (e.g. bringing a new product to market) generally occur within firms displaying one or more of the following features:

1. good external and internal communications: effective linkages with external sources of know-how; willingness to take on external ideas;
2. treating innovation as a corporate-wide task: effective functional integration; involving all departments in the project from its earliest stages; ability to design for ease of production;
3. careful planning and project control procedures: adequate up-front screening of new projects; regular project appraisal;
4. efficient development and high quality production: effective quality control procedures; using up-to-date production equipment;

[1] Roy Rothwell, 'Successful Industrial Innovation: Critical Factors for the 1990s', *R&D Management*, 22, 3, 1992.

5. strong market orientation: emphasis on satisfying user needs; efficient customer linkages; involving users in the development process, where possible;
6. providing a good technical service to customers, including customer training, where appropriate; efficient spares supply;
7. the presence of key individuals: product champions and technological gatekeepers;
8. high-quality management: dynamic open-minded managers; ability to attract and retain talented managers and researchers; a commitment to the development of human capital.

Rothwell finds that the factors associated with successful innovation are, more or less, common to all industries although their relative importance varies from one industry to another. There is no one single factor guaranteeing success. Successful innovators out-perform their rivals across the board.

Many of these factors for success emphasize communication – between functional divisions, with customers and suppliers, between key managers and product champions and other employees. Recently, John Kay of the London Business School has used the idea of the 'architecture' of the firm to examine how communication is established and maintained. There are three strands. *Internal architecture*, relationships between the firm and its employees and between employees; *external architecture*, relationships between the firm and its suppliers and customers; and *networks*, relationships between collaborating firms, all play a part in creating organizational knowledge, establishing a cooperative ethic and implementing and sustaining organizational routines.[2] Kay is in broad agreement with Rothwell when he notes the importance of architecture in securing successful exploitation of innovation, as distinct from mere invention.[3]

Rothwell neatly summarizes the importance of communication by concluding that innovation is essentially a 'people process' in which formal management procedures are no substitute for managerial talent and entrepreneurial flair. In short, innovation is a social activity.

---

[2] John Kay, *Foundations of Corporate Success: How Business Strategies Add Value*, Oxford University Press, Oxford, 1993, p. 68.

[3] Ibid., p. 110.

2.3. Innovation and Corporate Strategy

The eight project-related factors for success listed above and the concept of the firm's architecture describe what successful innovative firms do and how they organize themselves. Rothwell identifies these as *tactical* factors, as distinct from *strategic* factors. In doing so he is conforming to the dominant view that strategy is concerned solely with setting objectives: senior managers go on weekend retreats, find a vision of the future for their firm and then return to manage the inevitable process of corporate transformation. Kay is at pains to point out the shortcomings of this 'rationalist' view of strategy. On all past evidence, implementation of strategy is extremely difficult. New strategies which require major, rapid changes of behaviour in the firm tend to fail. Strategy must evolve in an incremental and adaptive manner. In this view, there is little sense in formulating strategic objectives without considering the existing organization of the firm. Strategies are 'emergent', being both inspired and circumscribed by the firm's existing strengths.[4]

A moment's reflection will reveal that, while it may at first seem counter-intuitive, corporate strategy which is emergent, evolutionary and incremental is better suited to innovation than the revolutionary variety. Revolutions necessarily disrupt the architecture of the firm during the change process and thus also interfere with the internal and external relationships on which innovation depends. In practice, Rothwell recognizes this in his taxonomy of *strategic* factors, again derived from observations of successful innovative firms. These strategic factors are:

1. top management commitment to, and visible support for, innovation;
2. a long-term corporate strategy in which innovation plays a key role;
3. long-term commitment to major projects;
4. corporate flexibility and responsiveness to change;
5. top management acceptance of risk;
6. the existence within the firm of a culture which accepts innovation and accommodates entrepreneurs.

---

[4] Ibid., p. 356.

The first three factors support the conclusion that successful innovation requires a stable corporate environment. The fourth is suggestive of adaptation and incremental change, rather than sudden and imposed changes in corporate direction.

Kay traces the idea of an emergent strategy back to L. E. Lindblom's writings on public policy in the 1950s. Lindblom asserted that political constraints on policy options ruled out a strictly rational approach to policy. For example, no UK industry official would suggest nationalization to a Conservative minister (not through fear for his job, but simply because there are more pleasant ways to waste one's time), even if this were a sensible policy option. To distinguish policy-making from rationalist decision-making, Lindblom coined the phrase 'the science of muddling through'. While some in government may take this rather too literally, Lindblom did himself an injustice, as his approach encompassed the idea that policy, although limited in scope, can be analysed, managed and controlled in a rational manner. So, too, can corporate strategy.

2.4. Innovation and Public Policy

Kay's observation that corporate strategy is limited in much the same way as public policy is particularly useful in leading us on to consider the interactions between environmental policy and innovation.

Governments create environmental markets. Unfortunately, many environmental policy-makers fail either to realize this or to recognize its implications for corporate strategy and innovation. Imagine a market analysis for some new type of audio equipment which resulted in the following conversation between the chief executive and the marketing department:

What size might the market be?

Either big, or zero.

When will demand begin?

Two to ten years, or never.

How will demand evolve over time?

Smoothly (growth, saturation, decline), or it might halt abruptly during early growth, or it might halt, resume, halt, resume....

How will the market vary geographically?

Each of our sales areas might display any of the above patterns.

What dictates these different scenarios?

Two hundred million people exercise a weak influence, every two years. A mixed bag of former lawyers, business people, teachers, minor public officials and media stars can kick the market into life. Any citizen or pressure group, a smart lawyer and a pedantic judge can kill it, at any time. Public employees throughout the country can depress the local market because they personally dislike or are unaware of the product. So when do we begin development, boss?

Er, I'll take a rain check on that.

Substitute 'pollution control equipment' for 'audio equipment' and you are in the weird and wonderful world of strategy formulation in the American environmental technology industry. Uncertain markets subject to unmanageable risks make the decision to pursue innovation more difficult to take. They bring additional risks where there is already likely to be a high degree of technical risk associated with the development effort.

Environmental policy has an equally important bearing on strategy and innovation in polluting firms which are the targets of policy instruments. Here it tends to be the production process rather than the output which is affected. The policy framework influences the competitive environment for the firm: managers need to decide whether they should put in the effort required to develop in-house technology and adapt external technologies to their specific needs, or simply resort to tried and tested methods. Uncertainty arising from environmental policy adds to the existing technical and organizational risks of technology development and adaptation. Doing more of the same old thing, i.e. not innovating, becomes more attractive.

Not only are firms in uncertain policy climates less likely to choose innovation in response to specific issues, they are also less likely to develop the architecture which provides continuous support for innovation. Architectures take time to evolve and adapt to new objectives. Firms will not embark upon the processes of internal change this requires if the reason for change – environmental policy – is inconsistent and unpredictable.

In a useful and wide-ranging review and analysis of how firms formulate and implement technology strategies, Metcalfe and Boden illuminate the link between a firm's organizational structure (architecture) and its areas of technological competence.[5] Technological competence, and the associated ability to innovate effectively, requires the presence of a community of technologists with a shared understanding of what is relevant to their area of expertise and of what constitutes a sensible course of action when designing new or modified products and processes. Old ways of thinking and working may become redundant as a result of a shift in the basic science underpinning the technology or of a change in the requirements of the market. When this happens the organizational structure must adapt accordingly.

Introducing environmental criteria often has the effect of creating an entirely new set of design criteria which may conflict with established beliefs about how to do things. When this happens the organization of the firm must adapt if it is to build up a technical understanding and capacity for sustained innovation which encompasses the new environmental objectives. The alternative to organizational change is to maintain the existing technological competence and rely on add-on solutions to meet the environmental objectives. This head-in-the-sand response favours expensive 'end-of-pipe' solutions over technically creative but potentially cheaper 'clean technology' ones. Unpredictable policy, as noted earlier, is less likely to inspire firms to make the effort to augment their technological competence.

Just as policy-makers may fail to appreciate their role in creating markets they may also misunderstand the dynamics of organizational change. Indeed, there is a risk that they may be misled by paying too much attention to fashionable consultants and management theorists who overstate the capacity for rapid, fundamental change within firms. As a consequence, some policy-makers fail to realize that by suddenly imposing environmental requirements, they do not allow firms sufficient time to adapt and so invite industry to pursue costly quick-fix solutions.

Almost as wasteful is the situation where a new environmental policy may initially inspire firms to restructure in order to pursue innovative, creative solutions, but fail to follow this up. In a normal commercial environment

---

[5] J. S. Metcalfe and M. Boden, 'Evolutionary Epistemology and the Nature of Technology Strategy', in Coombs, Saviotti and Walsh (eds), *Technological Change and Firm Strategies*, Academic Press, London, 1992.

there would generally be a constant incentive for innovation arising from pressures to reduce production costs or improve performance and features. In this way the initial effort in establishing new technical competence is rewarded by a continuous series of innovations arising from the changed technical expertise and organizational architecture. If environmental policy fails to set progressive goals this effort may be wasted, as the normal evolution of the technology cannot be played out. In the most wasteful circumstance the technical and organizational expertise may be lost, only for policy-makers to generate new demands many years down the line, when firms are again faced with the choice of adopting end-of-pipe solutions or making difficult organizational changes.

2.5. *Kaizen*: Continuous Improvement

A capacity to make continuous improvements in production is closely allied to, although not identical with, the facility for sustained innovation described above. Continuous improvement has been recognized as an important feature of the production process in certain internationally successful sectors of Japanese industry, especially car manufacture, from where it takes the name *kaizen*. It is dependent on a continuous, relentless pursuit of incremental improvement in all departments of the firm through, for example, quality circles, where every employee is encouraged to suggest ways to improve working practices and hence product quality. Opinions differ as to whether these improvements should be regarded as 'innovations' but they fit our broad definition of useful change within the firm, small though they may be when taken individually.

This all-pervasive attitude to quality and improvement in *kaizen* helps a firm to continually drive down production costs, through greater productivity, fewer errors and reduced scrap and wastes. *Kaizen* therefore complements low-cost efforts to reduce waste and emissions through 'good housekeeping', i.e. a focus on cutting out unnecessary pollution resulting from spills, sloppy work and unused raw materials. A culture of *kaizen* is undoubtedly accompanied by a certain type of organizational structure, but this element of the firm's structure tends to be stable and independent of particular sets of technologies employed within the firm. However, *kaizen* may be complementary with major shifts in technological competence which require organizational changes by making all employees comfortable with the idea of change.

2.6. Economic Theories of Policy and Innovation

Economic theory has been extensively applied to environmental policy but mostly in a limited fashion which fails to take adequate account of innovation. However, economic analysis which focuses on the effects of environmental policy instruments has in some countries been a major influence on policy-makers and has also strongly influenced their views on innovation. This section reviews what economics can really tell us about the links between policy and innovation.

Early contributions by economists, such as 'The Theory of Environmental Policy' by Baumol and Oates, 1975, were grounded in a static view of environmental policy which did not consider the effects of policies on innovation. They concentrated, for example, on calculating the level of regulation which would cause firms to reduce their pollution to what was in theory the socially optimal level. Even at this early stage some economists noted that advances in pollution control technology were likely to alter the equilibria of the standard analyses to such an extent that innovation should be the most important factor in determining environmental policy. However, models which consider innovation did not appear until the 1980s.[6]

At first, these models pointed to a common conclusion: that economic policy instruments such as taxes were not only less costly in the short run than traditional regulatory instruments such as emission standards but that they also encouraged more innovation. For example, when the theoretical groundwork for the Tradable Emission Permit (TEP), a type of economic instrument, was laid in the early 1980s, three main advantages were claimed for it: any overall level of emission reduction could be achieved at less cost than with a traditional regulation, the regulatory burden is reduced since firms are free to make their own decisions without reference to regulators

---

[6] See G. Ecchia and M. Mariotti, *A Survey on Environmental Policy: Technological Innovation and Strategic Issues*, Monograph 44.94, Fondazione ENI Enrico Mattei, Milan, 1994, for a helpful survey.

and, finally, each firm has an incentive to innovate by investing in more effective pollution abatement technology.[7]

This early consensus on the impact which TEPs have on innovation has since been overturned by Malueg, who used a more sophisticated model to show that TEP schemes can increase or decrease firms' incentives to innovate, depending on whether they are buyers or sellers of permits.[8]

In a wider-ranging theoretical study, Milliman and Prince also found that incentives for innovation are not necessarily improved by market-based instruments. They concluded that technology suppliers will be tempted to mislead regulators in unexpected ways. Under a tax regime, for example, the supplier will be inclined to exaggerate the cost of applying its technology, in the hope of constraining any tax reduction by the regulator and thus sustaining demand for the technology.[9] They also concluded that when an innovation reduces fixed costs (the cost of existing levels of abatement) rather than marginal costs (the cost of the next, incremental abatement), then polluting firms might favour upward ratcheting of standards when they are subject to all the types of regulatory instrument considered in the study, *except* emission taxes, when they would be opposed to higher standards.

Economic theorists have also tended to assume that firms and policy-makers have perfect information about the financial cost and the timing of innovations. Andrea Moro reviews some attempts to model more realistic situations where information is incomplete. Early results suggest that policy-

---

[7] W. A. Magat (ed.), *Reform of Environmental Regulation*, Balinger, Cambridge MA, 1982; T. H. Tietenberg, *Emissions Trading: An Exercise in Reforming Pollution Policy*, Resources for the Future, Inc., Washington DC, 1985; P. B. Downing and L. J. White, 'Innovation in Pollution Control', *Journal of Environmental Economics and Management*, 13, 1983, pp. 18–29.

[8] D. A. Malueg, 'Emission Credit Trading and the Incentive to Adopt New Pollution Abatement Technology', *Journal of Environmental Economics and Management*, 16, 1989, pp. 52–7. Malueg also demonstrates theoretically that the main claim for TEPs – that they achieve emission reductions at lower cost than non-economic regulatory instruments – has been overstated. In 'Welfare Consequences of Emission Credit Trading Programs', *Journal of Environmental Economics and Management*, 18, 1990, pp. 66–77, he examines the role of market structure. His model shows that in a situation where trading of permits is competitive, but the product market is dominated by only a few firms, a TEP system can reduce overall social welfare, primarily by reducing profits in the industry.

[9] S. R. Milliman and R. J. Prince, 'Firm Incentives to Promote Technological Change in Pollution Control', *Journal of Environmental Economics and Management*, 17, 1989, pp. 247–65.

makers should design complex, multi-part (i.e. flexible) policies, although Moro stresses the need for further research in this area.[10]

In general the results from the early economic models could be explained through simple deductive reasoning. Policy-makers, already used to receiving economic advice on the costs of regulations, were presented with convincing arguments for the superiority of one type of regulatory instrument over another as a tool for encouraging innovation. Recent models with more realistic assumptions have modified some of the earlier conclusions but their results are as yet less well known and are often more difficult to explain, leaving many policy-makers with the mistaken impression of a dichotomy between economic instruments (good for innovation) and more traditional regulatory instruments (bad for innovation).

Econometric models – theoretical descriptions of the effects of policies on an entire economy – have followed a similar evolutionary path, from concrete policy prescriptions to uncertain advice, as a direct consequence of attempts to assimilate innovation. Recent theoretical advances have been prompted by the debate over climate change. Policy-makers have a pressing need for advice on the economic consequences of taking action to reduce emissions of greenhouse gases, especially carbon dioxide. A series of econometric models developed around the time of the 1992 Earth Summit in Rio appeared to demonstrate that even if the costs of damage due to climate change are high, overall welfare costs are less if responses are delayed until there is greater scientific certainty about the existence and impacts of climate change. These models assumed that the costs of reducing emissions are fixed. However, by assuming that energy systems can adapt through technical change induced by the need to reduce emissions, Grubb, Chapuis and Duong recently concluded that delaying action on climate change can cost several times more than taking early action, even if the damage costs are modest.[11]

These inconsistencies suggest that, as far as innovation is concerned, economic modelling is immature and an uncertain foundation for policy

[10] Andrea Moro, *A Survey on R&D and Technological Innovation: Firms' Behaviour, Regulation and Pollution Control*, Monograph 73.93, Fondazione ENI Enrico Mattei, Milan, 1993.

[11] Michael Grubb, Thierry Chapuis and Minh Ha Duong, 'The Economics of Changing Course: Implications of adaptability and inertia for optimal climate policy', forthcoming in *Energy Policy*.

advice. In their wide ranging review of the treatment of innovation in environmental policy, Carraro and Siniscalco argue that: 'the main econometric models which are used in environmental policy analysis do not consider properly the determinants and effects of innovation, and offer unreliable estimates of policy effects', and 'There is much to be gained by merging the environmental literature with the industrial organization literature on innovation. The two fields have so far communicated little; but once the importance of innovation is recognized, the theory of industrial organization is the one to draw upon'.[12]

2.7. Summary

What are the important conditions for innovation which policy-makers should be concerned about? We need a more sophisticated viewpoint than simply 'economic instruments good; standards bad', or vice versa. This study approaches the interaction between environmental policy and innovation from a behavioural, organizational and strategic viewpoint. It is concerned with the relationships between regulators, industry and the public and the political and institutional aspects of environmental policy.

Policy-makers must be aware that they directly create the market for environmental technologies and the commercial environment which influences the decisions made by polluters. If they wish to create opportunity and incentives for firms to innovate, they need to understand the nature of decision-making in industry; how it is conditioned by perceptions of markets and risk, existing technological capacities and constraints and the capacity for change.

The effects of policy-making on innovation will be illuminated in the national studies and the two industry case studies. Major recurrent themes are the importance of policy stability and the timing and flexibility of policy instruments.

[12] Carlo Carraro and Domenico Siniscalco, *Environmental Policy Reconsidered: The Role of Technological Innovation*, Monograph 62.93, Fondazione ENI Enrico Mattei, Milan, 1993.

Part 2

Chapter 3

Denmark

3.1. Introduction

Denmark is particularly well suited to being the first in this series of six national studies. As the smallest country studied it offers a relatively simple policy environment, helped by the fact that Danish policy-makers have for many years adopted a consistent approach to environmental and industrial policy, and the interface between the two. The Danish study systematically outlines the main planks of Danish environmental policy. Since many of the key concepts in environmental policy recur in most developed countries, the subsequent studies can concentrate more on features which are unique to those countries.

The relationship between environmental and industrial policy in Denmark has been consciously managed for some time. In managing this policy interface, Denmark has developed specific mechanisms which encourage dialogue and understanding between parties who, elsewhere, might be adversaries. These mechanisms support sophisticated, flexible regulatory processes in which innovation can be supported and allowed to play a major role in meeting environmental goals.

3.2. The Development of Industry–Environment Policy

The current relationship between regulators and industry in Denmark has been shaped by two major forces: the cooperative, egalitarian principles inherent in Danish society and the political response to the great wave of environmental concern in industrialized countries in the late 1960s.

In response to growing public concern over the environment in the late 1960s, the government, at that time a left-wing coalition, set up a commission to investigate the extent of the problem being caused by pollution in Denmark, and make recommendations for government action. Before the commission had completed its task, a new right-wing government, led by the Social Democrats, came to power. Not wishing to be morally bound by the findings of the ongoing commission, and yet desiring to capitalize on the public concern for the environment, the new government moved quickly to establish a Ministry for Pollution. Following its own political instincts, the ministry opened a dialogue with industry (whose leaders were worried by the strength of environmental concern among politicians and public) on the form that an environmental protection regime for Denmark should take. This early dialogue with industry has developed into the cornerstone of present environmental policy.

The new Pollution Ministry began a process of consultation with Dansk Industri, the main representative body for Danish industry. The ministry wished to establish a system of regulating the most harmful industrial pollutants, which would take into account the technologies available for making improvements and the cost of these improvements. This was in essence a Best Available Technologies Not Entailing Excessive Cost (BATNEEC) approach. Most countries invoke concepts like BATNEEC, or BAT (without the cost element) in legislation applying to industry. In the United States about half a dozen related concepts have evolved in a succession of major items of environmental legislation.

It seemed natural in a decentralized country such as Denmark for the BATNEEC guidelines produced by the ministry to be applied by local authority officials familiar with the industries in their area, rather than by a national, centralized inspectorate. While industry was content with this, its leaders were less pleased with the proposed chain of authority for environmental regulations: any appeals against a local authority's interpretation of the ministry's guidelines were to be made to the Minister for Pollution, who would have the final say on the matter. Industry argued for an independent final arbiter, and in response a dedicated Appeals Court was established. This is composed of representatives of industry, ministry officials, environmentalists and legal experts. The establishment of the Appeals Court was a powerful symbol of the government's willingness to listen to the concerns of industry and, with hindsight, this early concession may be seen as a pointer to the level of trust which was subsequently to be built up between the two parties.

The general procedure established for setting environmental standards extended the principle of cooperation with industry. The Pollution Ministry's first priority was, and remains, to eliminate quickly any pollution damaging to human health. Beyond this, however, its personnel recognized the need for industry to adjust to new regulations and were concerned to bring all information to bear upon an issue before, for example, settling on an emission standard. Industry representatives were consulted at the earliest stages, whenever a new environmental issue arose. They participated, along with independent and government experts, in advisory committees which determined the potential for emissions reductions through existing and foreseen technology, and they were invited to participate in government-led research programmes aimed at developing the technical solutions needed to meet new standards, or to reduce the costs of achieving them.

Over time, it became clear that the government was serious about protection of the environment and that its general approach would be steadily to tighten standards. This in turn persuaded firms, which might initially have been reluctant, to cooperate with the government: clearly it is better for decisions on regulations to take account of a particular firm's own situation, and not just that of its competitors. This process rapidly reduced the 'asymmetry' of information between industry and regulator. Some of the mechanisms for achieving this are discussed more fully in Section 3.4 below.

There was also a wider political advantage flowing from the direct cooperation established between Dansk Industri and the Pollution Ministry. As in many other countries, industry's interests were well represented in parliament through Conservative MPs, who paid particularly close attention to the views of Dansk Industri. They were reassured by Dansk Industri's close involvement in laying the framework for Denmark's environmental legislation, the 1973 Pollution Act.[1] In addition the left-wing parties felt bound to support environmental policy, even though they may have had some concerns that industry's voice would be too strong. This concern was most vocal among grass-roots environmental groups. However, these groups, and their successors, have largely been won over to the Danish system by

[1] The primary legislation on the environment is now the Environmental Protection Act, which entered into force on 1 January 1992.

the consistency of the government's approach and the formal role they are given in advisory bodies at all levels.

3.3. Current Environmental Policy

Denmark provides a relatively straightforward illustration of the main elements of environmental policy, most of which recur in other industrialized countries. It is worth while taking some time here to set out those general areas of environmental policy where industry is most directly affected. This will provide some useful background to draw upon in some of the later chapters.

Danish environmental policy is the responsibility of the Environment Ministry. Since late 1994, this responsibility includes energy policy. The former Energy Ministry had responsibility for energy policy, but increasingly this had become subsumed into a framework where environmental issues were of predominant importance. It is worth while to examine both policy areas in turn.

3.3.1. Environmental Policy: The Main Elements

The Environmental Protection Agency (EPA) of the Environment Ministry (the successor to the original Pollution Ministry) identifies five main areas of activity.

- Approval of operations, under the BATNEEC regime

This is the EPA's largest single activity, in terms of administrative effort. The EPA is responsible for identifying polluting or hazardous activities, producing guidelines (regulations) for dealing with these activities and assigning responsibility for applying the guidelines to other public authorities.[2]

The EPA negotiates the guidelines with industry in advance, through industrial organizations, but also consults formally with other interested parties. The agency extends an open invitation to any organization which wishes to be consulted on proposed BATNEEC guidelines. Currently there are 57 organizations, including many environmental groups, on the official consultation list.

---

[2] The Environmental Protection Act outlines the duties and powers of the Environment Minister.

Expert working groups play a central role in this process, by determining what is technically achievable. The key organization in this process is the Danish Technology Institute (DTI), an independent industrial research organization which has traditionally received the bulk of its funding from Industry Ministry and Environment Ministry programmes. In recent years, the Environment Ministry has provided by far the biggest proportion of the DTI's funding. As a result, it has built up a detailed and comprehensive knowledge of the technology base of Danish industry, and of the existing and prospective environmental technologies. The DTI generally has as good a knowledge as the industry of what emissions reductions can be achieved through improved technology, over what timescale and at what cost. It can also very rapidly devise an appropriate joint R&D programme with leading firms and monitor progress towards the required technical and economic standards.

In most cases, the EPA's development programmes provide support for a full-scale demonstration of the new technology. This is seen as essential for establishing that it meets the requirements of "A" (available) and NEEC (not entailing excessive cost), and that local authority regulators will have confidence in citing it as a benchmark, when granting approvals for new processes.

Once guidelines for specific activities have been approved, local authorities generally have responsibility for implementing them. Local authority officials inspect individual firms to ensure they meet the requirements of the guidelines and then issue approvals or licences for the polluting activities. These local officials form Denmark's industrial pollution inspectorate.

• Support for cleaner technologies

The programme of support for the development of cleaner technologies and processes was motivated by the EPA's attempts to regulate emissions of heavy metals, where it became clear that due to the great variation in polluting processes end-of-pipe solutions would be very costly. The clean technology programme is focused on developing solutions to current pollution problems, reducing the costs of complying with existing emissions standards, and achieving the timetables for tightening standards set out by the EPA. The bulk of the funds flows through the DTI, with some being spent within that organization, and contributes to joint programmes with industry.

• Agreements with industry

This is a new area of activity in Denmark, where no experience of success or failure yet exists. The agreements are modelled on the voluntary agreements in the Netherlands (see Chapter 4) and their implementation in Denmark can best be understood by looking at the agreement which has been developed furthest: that of Volatile Organic Compounds (VOCs). These chemicals are used in a wide variety of products such as solvents, paints and coatings in industries ranging from the largest manufacturing concerns to the smallest paint-shops, and include evaporative emissions of vehicle fuels such as gasoline.

Under a UNECE (United Nations Economic Commission for Europe) Protocol, Denmark was expected to reduce emissions of VOCs to 30% below 1988 levels, by the year 2000. The EPA then announced that it wished to pursue a target of a 50% reduction by the same time. However, the traditional thorough consultation process, by which the practicality and costs of meeting the target could be established, was not appropriate for a substance with such a huge variety of uses and users. The EPA therefore proposed a blanket 50% target for all uses of VOCs, to be enforced through the existing industrial licensing system.

Industry associations, notably Dansk Industri, protested that some firms might be forced out of business by such a target, while others could easily achieve far more. Dansk Industri estimated that the cost to industry would be 1.2 billion krone ($180 million). A traditional reaction might have been for the regulator to allow exemptions for some industries. This is usually a messy solution, where firms see advantage in exaggerating their costs and difficulties and other more deserving cases fail to get a fair hearing. Instead, the EPA asked Dansk Industri to devise a programme of reductions, based on best offers from the organization's member firms. The EPA gave no guarantees that it would be satisfied with the voluntary offer, thus imposing a strong incentive on firms to make a maximum offer from the outset.

Dansk Industri is organized into branches representing individual industries. Each branch was asked to produce an overall reduction target, based on members' views of what could reasonably be achieved by developing and applying clean technologies and processes. During the discussions with Dansk Industri, individual firms did not know what the overall branch target or the final aggregated offer would be. In the event the aggregated target offered by Dansk Industri was a 40–45% reduction, within which some branches have targets of, for example, 70%, and others only 20%. Every firm has a specific target, and it is hoped that poorer than expected progress in some firms and branches will be compensated by better progress elsewhere.

Firms wishing to participate in the agreement must join Dansk Industri. Those who do not take part will be regulated by the existing licensing regime, where the EU target of a 30% reduction will be a minimum requirement for all non-agreement firms. The local authority will rapidly adopt the best standards being achieved by members of the agreement as the norm for licensing purposes. The non-member firms will be at a disadvantage, in some cases because their competitors will have targets of less than 30% under the agreement, and in all cases because they will not have been involved in developing the know-how required to meet the improved performance standards demanded by regulators making licensing decisions. Many observers in Denmark expect a rapid increase in Dansk Industri's membership as a result of this pressure.

The Danes do not refer to this mechanism as a voluntary agreement, because all sides recognize that it is nothing of the sort. It is a target agreed under duress, where the regulator reserves the right to impose stricter standards if it becomes clear that the target is unlikely to be met. Background threats of this nature are characteristic features of voluntary agreements, being used extensively in the Netherlands and in the long-established agreements in Japan.

• Economic incentives

Denmark makes extensive use of pollution charges and fees, especially for waste treatment and disposal. The EPA believes that these economic instruments have been effective in reducing the amount of wastes being generated by industry. However, economic instruments are regarded with suspicion and resented by industry. Its main objection is that pollution charges, such as waste water treatment taxes, deprive firms of the money they require to develop and invest in cleaner processes. Another common objection is that the charges often represent only a small percentage of the cost of the resource or service, and so provide no realistic incentive for innovation.

Industry suspects that pollution charges are being used simply as a means of raising revenue. While some of the government's revenue from pollution charges funds development and demonstration of clean technologies, the majority goes to the Finance Ministry. As an umbrella organization concerned with equalizing the impacts of government policy on its members, Dansk Industri also objects that only a minority of firms benefit from demonstration programmes which do happen to be funded from pollution charges, and often these beneficiaries are third-party equipment manufacturers or consultants, not the users who pay the taxes. The organisation accepts that economic instruments can be an effective means of raising the funding required to develop clean technologies, if all the money is diverted to this purpose, and that this might provide incentives to innovative firms, but it points out that this has not been the experience in Denmark.

Recently, the Finance Ministry has proposed a hefty tax on the time that local authority officials spend on processing firms' applications for licences for polluting activities and on providing advice to firms. This perverse measure threatens to impose arbitrary (and in some cases, terminal) costs on firms, to destroy the hard-won dialogue between regulators, environmentalists and industry and to arrest the diffusion of clean technologies and processes. It underlines the need for regulatory policy – whether this concerns traditional forms of regulation such as limits on emissions or newer forms such as economic instruments – to be conducted by knowledgeable officials with an understanding of the impacts they are likely to have, or at least a good working relationship with the industries concerned.

- Provision of infrastructure

The final area of EPA activity concerns the agency's duty to ensure that adequate infrastructure exists in Denmark to cope with the needs of industry in relation to waste treatment and disposal. To meet this responsibility, the EPA has pursued a policy of centralizing waste treatment facilities for industry. Local communities fund the provision of these facilities, through their municipal authorities, and firms are obliged to use them for all their waste treatment and waste disposal needs.

The EPA is now attempting to make its environmental policy more coherent and comprehensive, in keeping with its wish to make sustainable development into a concrete policy objective for Denmark. It is planning a national audit of the way in which all primary materials flow through the Danish economy. Full life-cycle analyses will identify those materials which have the greatest impact on the environment and these will become the EPA's priorities for future action. This is both a top-down approach (inspired by the question 'What harms the environment?') and a bottom-up approach (tracking each material through society), and as such represents a break from the reactive environmental policy which has been the norm in Denmark, and elsewhere, since the surge of environmental awareness in the 1960s. This kind of approach may be an essential element of a credible policy for sustainable development, in any country.

3.3.2. Energy and Environmental Policy

Even more than in other developed countries, environmental objectives are a major feature of Danish energy policy, although security of energy supply remains an important concern. The policy document *Energy 2000* lays out a detailed plan of action for achieving environmental goals.[3] Danish energy policy has the same strong cross-party support that environmental policy enjoys, giving the former Energy Ministry a great deal of persuasive power over industry. For example, utilities have recently adopted measures to reduce consumer demand. However, they did not do this of their own accord, and indeed they were reluctant to become involved in demand-side management measures. The Energy Ministry threatened to invoke the demand-side management goals mentioned in *Energy 2000*, and formally introduce measures to force the utilities down this route. The result was, according to the Ministry, a culture change in the utilities, which are now voluntarily encouraging consumers to reduce their energy use.

In addition, the Energy Ministry had a powerful planning function, retained by the Environment Ministry, allowing it to issue directives to regional authorities on matters such as the proportion of the population which must be supplied by District Heating by a certain date, the fuel to be used for this purpose and the generating mix for electricity supplies. Thus, the ministry ensured that markets existed for clean energy technologies, notably wind energy and biomass fuels, by requiring that local authorities contract specified amounts of power generation from these sources. In the case of renewable energies this has been backed up by generous funding of R&D for Danish manufacturers. Officials of the former ministry recognized that some of this

[3] *Energy 2000: A Plan of Action for Sustainable Development*, Danish Energy Ministry, April 1990.

funding was a thinly disguised subsidy for wind turbine manufacturers, who were suffering the effects of over-expansion based on the prospect of export markets that failed to materialize.

Like the EPA, the Energy Ministry stressed the importance of regulators being well informed about the potential for improvements in technologies and practices, through expert working groups and government-supported research. It also identified the expertise of ministry officials, many of whom have worked in energy-related industries, as contributing to the ministry's ability to block attempts by firms to mislead it.

Given the predominance of environmental objectives in energy policy it is perhaps unsurprising that energy responsibilities were taken over by the Environment Minister, even though this may be unique among OECD countries where energy policy, if it is not separate, tends to be run by Industry or Economy ministries.

3.4. Fostering Innovation

It should be clear from the above discussion that Danish regulators are concerned that industry should be allowed time to adapt to regulations, without compromising the original environmental objectives. They assume from the outset that innovation will play an important role in meeting those objectives, and are prepared to support the research and demonstration projects which are often essential to this process.

At its best, the Danish approach to environmental regulation is self-regulating and relatively free of political pressures. A new environmental problem will spark off a process of negotiation between government and industry on the best, practical approach to the problem. This open regulatory process has been successful in setting standards which are demanding and yet realistic. One key to this success has been the reduction of the asymmetry of information between regulators and industry. This has been achieved in the following ways:

1. Knowledge Capture

The government has built and maintained knowledge of the state of the art in environmental technologies and techniques within a limited number of government or quasi-government technical institutions. This is achieved through consistent funding of research, development and demonstration programmes, channelled through organizations such as the Danish Technological Institute, which are themselves involved in the practical aspects of the work. The programmes include industrial funding and involvement, but the ratio of public to private funding is relatively unimportant, so long as the public funding is sufficient to attract the genuine involvement of industry.

Another process assisting knowledge capture is the ease with which individuals move between the public and private sectors. This ensures that regulators, even at a senior level, often have experience of the industries they are regulating. In many cases the individuals in the public and private bodies involved in a regulatory process will have been colleagues in the past. These personal relationships further reduce the likelihood that firms will attempt to mislead regulators.

2. Credibility

It is one thing for firms to become involved in publicly supported programmes, but quite another for them willingly to divulge, from the outset, the information which allows the regulators to set sensible environmental objectives and devise appropriate supporting R&D programmes. When a particular firm's current product line or working practices are threatened by new regulatory proposals it has two choices. It can elect to cooperate, and work with regulators to establish realistic targets for improved environmental performance or it can fight, by refusing to divulge its own knowledge, promulgating misleading information and complaining loudly about the costs of proposed measures, in the hope that the regulators will back down.

Danish firms know from their own experience over the last 20 years that the regulator does not back down. In fact, the EPA has demonstrated that it will set tight standards in the face of industrial opposition, where it feels that cooperation has not been forthcoming from industry. The situation has now been reached in Denmark, where firms know that they will generally be given a fair, informed hearing by knowledgeable regulators and that the best way to protect their future business prospects is to participate openly and at an early stage in negotiations on new regulations, and to maintain where possible a role in the development of the state of the art on which future regulations will be based.

3. Institutional and Political Power

The Environment Ministry's ability to pursue a credible and consistent policy over the last 20 years is based on its extraordinary degree of power within the Danish political system. The first part of this chapter described how the foundation of this power was laid in the mid-1970s when the new Pollution Ministry forged a strategic alliance with industry, thus cutting the Industry Ministry out of any significant role in the environment–industry policy area. One result has been that the EPA's substantial technology budget has been exempt from recent government spending cuts, while the Industry Ministry's budget has been sharply reduced. The EPA is now working this trick on the Foreign Ministry, with respect to aid programmes for developing countries, as described later in this chapter. The Environment Ministry's takeover of energy policy is the latest confirmation of its institutional and political power.

The Environment Ministry has been careful to respect its side of the early bargain with industry, involving firms in the formulation of environmental regulations and generally avoiding confrontations. Industry recognizes that the Environment Ministry is now sufficiently powerful to pursue its objectives regardless of industry's objections. The only sensible response for a firm is to cooperate fully with the regulators, divulge all relevant information and trust that the outcome will be in keeping with its own technical capabilities.

3.5. Two Cautionary Tales

On occasion the Environment Ministry has not adhered to this successful model of bringing all relevant information to bear on an issue and building a consensus around the measures to be taken. Two recent examples illustrate how even the best environmental regimes can be derailed by sudden and intense public concern, or by politicians pursuing policies for symbolic, rather than environmental, purposes.

3.5.1. Pipe Dreams

The government's approach to ozone-destroying chemicals is one example of an environmental goal being transformed into a political totem. The Environment Ministry is determined that Denmark should be the first industrialized country to eliminate the use of CFCs and their less damaging replacements, HCFCs. To achieve this the ministry is going far beyond the requirements of the Montreal Protocol and its amendments, which set the phase-out timetable for more than 100 signatory nations. Denmark's action will have a negligible effect on global emissions, but is imposing a large cost on its domestic industry.

For example, in 1988 manufacturers of District Heating pipes were forced to replace CFCs with HCFCs in the production of the insulating material used in the pipes. Shortly after the manufacturers had completed the required changes in their production processes, the government announced that they must replace HCFCs with chemicals which cause no harm to the ozone layer (pentane and CO_2 are the industry's preferences) by the end of 1993, even though the Montreal Protocol currently allows production of HCFCs up to 2015.

The cost of accommodating two changes of technology within a short space of time has been high and in the case of the HCFC process there has been insufficient time to recoup any of the development and investment costs. Cheaper pipes produced in other EU countries using HCFCs threaten the domestic market and the government has responded by proposing a ban on their import. This has resulted in a legal battle with the European Commission, which claims that Denmark is breaching the rules governing the European Single Market. In reply, Denmark has attempted to claim the environmental moral high ground.

3.5.2. Fishy Business

When fish deaths in several small rivers were attributed to textile factories discharging chemicals used in their colour dyeing processes, the government's response was, according to Dansk Industri's interpretation, hasty and provoked by the clamour in the media.

The EPA responded to the public outcry by establishing strict guidelines on discharges of methyls, the chemicals causing the damage, from all textile firms. These guidelines protected the most sensitive ecosystems, i.e. small, freshwater rivers, but made no concessions to firms which discharged straight into the sea, where rapid dilution occurs and no threat to marine organisms had been established.

Dansk Industri points to the fact that the normal consultation procedures were bypassed in this case, and blames these regulations for the decision by many textile firms to close their Danish factories and relocate in the Baltic states. However, this interpretation must be set against the general trend for the textile industry to abandon Western countries in favour of those with lower wage costs. It is difficult to determine whether the regulations caused a significant increase in costs, acted as a catalyst for relocations which were being considered, or were simply an excuse for decisions which had been taken for other reasons.

Episodes such as these clearly illustrate how easy it may be for strains to develop within a culture of cooperation between industry and regulators which has been nurtured over many years. In Denmark, the basis for this cooperation has been a trade-off in which firms give up the power which comes from their superior knowledge of technologies and markets in exchange for a seat at the regulatory negotiating table.

Danish regulators have not considered to any great extent the ways in which their regulatory system affects firms' strategic incentives for innovation. In part, this is because both industry and regulators assume that innovation will play a significant role in meeting new standards, and plan for this from the outset. Regulators seek to ensure that Danish standards are in general on a par with the strictest standards encountered in other countries. This ensures that Danish exporters have little difficulty in meeting the standards required in any foreign market. The EPA even admits to an underlying ambition for Danish standards to lead the world, but this is based on a belief that standards elsewhere are then more likely to be modelled on those in Denmark, and will therefore be easily met by Danish products. The prime motivation is to avoid the possibility of foreign standards becoming a de facto trade barrier to Danish products, rather than stimulating a technological first-mover advantage through high standards.

3.6. Global Ambitions

Danish industry is export-oriented. As Denmark has few natural resources, it must rely on exporting a high proportion of its industrial output in order to support a high standard of living. However, as a relatively small country it has a scarcity of human resources and so must concentrate on doing a small number of things well. Danish firms have become flexible in order to meet the demands of varied export markets and they are used to making rapid and frequent changes in production.

Denmark presents a remarkable example of partnership between industry and government in harnessing the growing worldwide concern for the environment as a vehicle for promoting Danish exports. In essence, Denmark is setting out systematically to export its environmental standards to industrializing nations (and nations undergoing post-communist restructuring). This will in turn create markets for the technologies which Danish firms developed in response to domestic environmental standards. If it succeeds, the Danish strategy will stand on its head the prevailing wisdom on trade and the environment, i.e. that developing countries threaten to destroy Western industries with low-cost goods produced by unregulated dirty industries. Denmark's approach is illustrated by a specific example of regulatory capture, and a proposed aid-for-regulations development programme which will consume a large proportion of the Danish government's overseas aid budget.

Denmark has 50% of the world market for the pipes used in District Heating systems, and exports 80% of its production. The pipes carry high temperature water from central boilers to domestic and industrial users. In countries with realistic energy prices, minimizing the heat lost in transporting the water through the network of pipes is a priority, so the pipes are generally insulated to a very high standard. This insulation is traditionally made of expanded polystyrene foam produced using CFCs – the main culprit in the destruction of the ozone layer.

In the restructuring economies of Eastern Europe and the former Soviet Union the use of District Heating was widespread. Energy was heavily subsidized and domestic users frequently paid nothing for heating supplied through the District Heating networks. Domestic radiators fed by the networks rarely included temperature controls and typically could not be turned off. Room temperature was regulated by opening windows. The systems were badly maintained and prone to leaks and the pipes were generally uninsulated. As a result the heat losses were so great that in the coldest weather those furthest from the boiler plants were supplied with water at too low a temperature to be of any use. Meanwhile the boilers were working at peak capacity and very high output temperatures, causing problems with overheating for those closest to the plants. Western observers and aid agencies quickly recognized that in some of these countries the dilapidated state of the District Heating systems was a major contributor to social hardship.

The World Bank is an important source of support for restructuring projects in Central and Eastern Europe (CEE). Lending for energy infrastructure projects has traditionally been a focus of World Bank activity and after the fall of communism it was preparing to support projects aimed at improving the energy efficiency of the District Heating networks in several CEE countries. As described above, Danish companies had already been required to switch production from CFC-based insulation to the less damaging HCFC-based insulation. Danish companies and the Danish government were trying to persuade the World Bank to require all projects seeking World Bank support to use pipes with HCFC-based insulation.

World Bank officials were sympathetic to the environmental argument and there was little difference in cost between the two types of insulation. However, they placed a high priority on the reliability of any new networks and were concerned that there had been little experience with the Danish HCFC-based insulation. The Danish Environment Minister was sufficiently convinced of the importance of favourable World Bank regulations to Danish industry that he personally invited the relevant officials to Denmark to see for themselves the production facilities of the Danish companies and the demonstration projects funded by the government. The World Bank officials were reassured and promptly adopted HCFC-based insulation as a standard for all District Heating projects with World Bank involvement, delivering a significant advantage to Danish firms in the large CEE market.

It is likely that the Danish overseas development effort will in future be dominated by environmental objectives aimed at introducing clean technologies in developing countries, and (not coincidentally) boosting Danish exports. The proposed Environmental and Disaster Fund will consist of two elements: the existing fund for disaster relief in developing countries and a new fund for environmental aid. The fund is projected to consume 0.5% of Danish GNP – 4.4 billion krone – by the year 2001, building up from 1.1 billion krone in its first year of operation, probably 1995. The greatest part by far of the budget will be earmarked for the environmental part of the Fund.

The beneficiaries will be the countries of Central and Eastern Europe and developing countries such as Venezuela and Malaysia which already have an appreciable industrial infrastructure. The fund will support a range of measures in these countries to encourage the introduction of environmentally friendly technologies, but most of the budget will be used to support demonstrations of Danish technologies. This hardware effort will be supported by education and awareness-raising, strengthening of institutions and advice and assistance with introducing environmental regulatory regimes, which will, of course, underpin the standards achieved by Danish technologies in the demonstration projects.

There are two selfish justifications for the new fund. First, it is a response to those Danish firms which complain that they get little benefit, in terms of contracts, from the Danish government's contributions to multilateral aid agencies such as the United Nations Development Programme. Small countries generally feel that these agencies are dominated by the large donor countries, and that contracts for development projects are disproportionately awarded to firms from these countries. The new Danish fund is likely to be partly funded by reductions in Denmark's contributions to such multilateral organizations. Secondly, the combination of successful technical demonstrations and familiar regulatory regimes should create new global markets for the clean technologies which Danish firms have developed in response to domestic regulations. The decision to focus on countries with prospects for rapid short-term industrial growth is justifiable on purely environmental grounds – more industry means more potential pollution, which can be avoided with clean technologies – but also makes sense in the context of a desire to maximize the size of the new markets for Danish technologies.

The proposed fund also provides a further example of the increasing power of the Environment Ministry and the EPA within the Danish government. The Foreign Ministry is currently responsible for all foreign aid programmes and is incensed at the prospect of the EPA taking over such a large part of the overall Danish aid budget. However, the EPA points out that the Foreign Ministry has no experience of managing advanced technical projects. More importantly, the EPA has secured the support of Finance Ministry officials by promising to give them a significant degree of control over the programme – something which the Foreign Ministry has always denied them.

3.7. Summary

The clear structure of environmental policy in Denmark helps us to identify those factors which are important in allowing innovation to play a role in the policy and regulatory process, and ultimately to provide solutions to environmental challenges.

First, there is the overarching political context of the environmental issue in Denmark. An early fusion of interests took much of the political heat out of environmental issues. This was built upon a generally corporatist political landscape in Denmark, where many types of interest group are drawn into decision making. Even so, environmental issues of the late 1960s and early 1970s had the potential to disrupt this consensual tradition. Politically, the solution lay in elucidating a clear environmental policy and sticking to it, resisting (for the most part) temptations to pander to those supporting or opposing environmentalists.

At the same time, policy-making at all levels was embedded in dialogue mechanisms which included industry and environmental groups, other citizens' groups and independent experts. As a result environmental and industry groups were less inclined to take adversarial positions, and this limited the political strains on national environmental policy. Further detailed, technical dialogue between government and industry occurred largely under the auspices of the Danish Technology Institute. This multi-level dialogue with industry brings out essential technical and commercial information and allows regulatory targets to reflect, anticipate and respond to innovation.

The Danish example indicates that the exact legal definitions of best available technology, state of the art, maximum achievable control, and related concepts, have little bearing on the prospects for innovation to flourish within a regulatory regime. The dialogue processes available to policy-makers and the political context within which they occur are the key to successfully harnessing innovation.

Chapter 4

The Netherlands

4.1. Introduction

As in many other countries, environmental issues came to the fore in the Netherlands during the 1960s. The new environmental concerns reflected the unique situation in the Netherlands. As a small, densely populated country, situated largely in the delta of the Rhine, attention was focused on issues such as surface water pollution and waste disposal. This new environmental consciousness collided head-on with the effort to rebuild the economy after the ravages of the Second World War. To this end, the Dutch government's over-riding priority had been to promote intensive agriculture and the chemical, iron and steel industries, and by the late 1960s this policy had delivered reasonable prosperity.

Public perceptions of the ideal balance between growth and the environment had shifted noticeably by the time *The Limits to Growth* was published in 1972.[1] With its high population density, the Netherlands seemed to be a living example of the book's dire warnings of environmental disaster. Around this time various laws, such as the Pollution of Surface Water Act (1970) and the Air Pollution Act (1972), were created to deal primarily with overt, local pollution. They adopted the traditional approach of imposing standards on a reluctant industrial sector.

In the 1980s a series of new environmental problems jolted public and politicians into a realization that, while visible local pollution had been restrained, diffuse, widespread pollution had grown unchecked. Nitrate pollution from agriculture was poisoning groundwater, acid rain was harming ecosystems in the Netherlands and abroad, and chemical waste and other dangerous industrial waste was accumulating in the soil. Out of this grew a widespread commitment to sustainable development.

---

[1] D. L. Meadows, *The Limits to Growth*, Universe Books, New York, 1972.

The result was the world's first comprehensive national plan for a sustainable economy: the National Environmental Policy Plan (NEPP). This establishes a new model for environmental policy, in which the relationship between regulator and industry has been redefined as a cooperative one, working towards a mutually accepted goal – an environmentally sustainable economy. Much of this chapter will be devoted to the workings of the NEPP, its likely impact on innovation and its ambitious aim of encouraging a radical transformation of industry over the next 20 years.

4.2. The National Environmental Policy Plan

The National Institute of Public Health and Environmental Protection surveyed the state of the Dutch environment and published the results in the report *Concern for Tomorrow*.[2] Inspired by the Brundtland Commission Report on sustainable development (*Our Common Future*),[3] *Concern for Tomorrow* concluded that emissions of many industrial pollutants would need to be reduced by 70 to 90% by 2010, if environmental disaster was to be avoided.

This was a far greater reduction than anything achieved by traditional legislative instruments. In recognition of this the government began to talk openly about the need for 'internalization of the environmental problem'. More than simple internalization of the social costs of pollution, this approach was based on fostering environmental sensitivity among all citizens, administrative bodies and especially industry, each taking responsibility for helping to solve environmental problems. An all-party political consensus and broad public support for this approach was soon achieved and the Environment Ministry took on the task of devising an implementation plan.

The Dutch government's guide to the National Environmental Policy Plan sets out the rationale for this revolutionary piece of legislation.[4] The preamble clearly states that the NEPP is a response to the need to move towards

---

[2] *Concern for Tomorrow*, RIVM, Bilthoven, 1989.

[3] World Commission on Environment and Development, *Our Common Future*, Oxford University Press, Oxford, 1987.

[4] Ministry of Housing, Physical Planning and the Environment, *Highlights of the Dutch National Environmental Policy Plan*, VROM 90678/1-90, 1990.

Table 4.1: The Five-Level Model in the Dutch NEPP

Scale	Typical problems	Major goals of NEPP
LOCAL the developed environment	Indoor Environment Soil	Sharp reductions in noise and odour
REGIONAL the landscape	Eutrophication Waste Disposal	Reduce acid emissions and waste by 70–90%
FLUVIAL river basins and coastal seas	Eutrophication Deforestation	Reduce eutrophying and non- degradable substances by 90%
CONTINENTAL air and ocean currents	Acidification Fall-out	Reduce acid emissions and some hydrocarbons by 90%
GLOBAL the upper atmosphere	Ozone Layer Depletion Climate Change	Stabilize CO_2 emissions at 1989/90 levels by 2000

sustainable development. All groups and sectors of society are expected to be involved in this effort.

4.2.1. Structure of the NEPP

The intellectual and scientific underpinning of the NEPP is provided by *Concern for Tomorrow* which presents the environment as a system of reservoirs, with natural cycles of substances circulating within and between these reservoirs. Five levels of scale are distinguished, ranging from the local to the global level, each with its own environmental problems (see Table 4.1). Problems at one level have effects at higher and lower levels, so that global warming, for example, affects the local environment through the damage caused by extreme weather events. 'Diffusion' is the movement of pollutants from lower to higher levels, causing environmental problems at each stage. At higher levels, problems take an increasing length of time to become apparent and counter-measures become progressively difficult and slow to take effect.

The NEPP itself is a strategic long-term plan (1990–2010) setting out the broad environmental policies necessary to achieve sustainable development by 2010. It adheres to the following set of principles: polluter pays; abatement at source; stand-still (no further deterioration of environment); application of best practical means of abatement; control of waste disposal; internalization (encouraging environmental values in individuals).

With these principles in mind, there are three processes for achieving the long-term goals:

- closing substance cycles, also referred to as 'Integral sequence management', in which the aim is consciously to manage the entire chain of production, consumption and disposal, to maximize re-use and recycling and minimize emissions at each stage;
- conserving energy and using cleaner energy sources;
- quality enhancement, i.e. promoting the highest quality of production processes and products.

At each scale the NEPP sets out the major goals to be achieved by 2010, or sooner (see Table 4.1). Other goals – which are closer to being aspirations – for all countries are also suggested, e.g. to stabilize the concentration of greenhouse gases in the atmosphere by 2010.

Emissions abatement is to be followed in one of the following three ways (in ascending order of preference):

- *emission-oriented:* reduce emissions and waste through end-of-pipe clean-up techniques;
- *volume-oriented:* use policy measures to reduce the scale of production;
- *structure-oriented:* employ more economical and/or cleaner production and consumption processes.

Initially, the emphasis will be on emission-oriented methods but to meet the NEPP targets there will need to be increasing emphasis on volume-oriented methods up to 2010. Structure-oriented methods are expected to have an impact only over the longer term because of long development lead-times. Widespread application of structure-oriented methods would lead to a radical transformation of the Dutch economy. Uptake is constrained by the length of typical investment cycles for large-scale developments and by the speed at which structural changes, analogous to earlier shifts away from heavy industry, can take place without excessive social upheaval.

Several initiatives aim to explore the changes in design and materials which are necessary for structure-oriented methods to succeed. The Ministry of Economic Affairs' research programme on environmental technologies is addressing eco-design for waste minimization. This includes an assessment (but not a complete life-cycle analysis) of the environmental impacts of a range of primary manufacturing materials. The results will be made available to manufacturers, in the hope that they will include the environmental impact of materials in their design criteria.

4.2.2. Target Groups

The NEPP identifies ten target groups within Dutch society. Each of these groups is expected to make its own contribution towards meeting the major goals of the NEPP. The target groups are:

> agriculture; traffic and transport; industry (including refineries); energy; construction; waste processing; water supply; environmental equipment manufacturers; research institutes; societal organizations; and consumers.[5]

Just how the major goals of the NEPP will be distributed among the target groups remains partly unresolved. Progress has been easiest in those areas where it is clear that the national problem arises from the activities of only one target group. For example, in agriculture, pesticide use is to be reduced by 50% by 2000. For more complex problems, where the proper distribution between target groups is less clear, much of the effort to date has been directed towards setting broad targets for the industry target group. Discussions between the government and industrial organizations, especially the Netherlands Employers' Organization (VNO), backed up by consultants'

---

[5] 'Societal organizations' includes employers' and employees' organizations, youth and senior citizen groups and 'ideological' organizations. Environmental organizations are to have an 'antenna' function and will be given a major role in contributing ideas for structural changes to society and the economy. Even the role of the police has been addressed: they are to ensure effective enforcement of the NEPP, where necessary, and to this end a regional administrative structure and programme of environmental training have been created within the police force. Consumers become an explicit target of consciousness-raising and will, for example, be expected to separate all household waste into classes such as plastic, organic and paper, by the year 2000.

reports, produced agreement in principle on a range of targets. For example, SO_2, NOx and VOC emissions must be reduced by 80%, 45% and 45–60% respectively, by 2010.

Closely linked to the discussions on targets for industry were parallel discussions on the appropriate mechanisms for achieving these targets. Government and industry agreed that attempting to achieve reductions of the desired scale through traditional forms of regulation, e.g. imposing the same emission reduction requirements on all industrial processes, would be extremely costly and inefficient. An ideal mechanism would take account of differences between industrial sectors in their ability to make reductions and even differences between firms within the same sector. The solution, which was instrumental in gaining VNO's acceptance of the NEPP, was the system of voluntary agreements between the government and industrial sectors.

4.3. Voluntary Agreements with Industry

By the end of 1993, around 15 voluntary agreements had been concluded with different groups. Most of these groups are associations of firms within narrowly defined industrial sectors, such as the packaging, fine chemicals and base metal industries. However, agreements also exist with sub-sectors of other target groups such as agriculture.

Several pressures contributed to the adoption of voluntary agreements as the key mechanism for implementing the NEPP within the industrial sector. As noted above, both parties recognized the advantages they conferred in terms of flexibility between sectors and individual firms to trade off reductions in emissions. However, industry was very keen to use the agreements as a means of eliminating the uncertainty associated with possible future environmental targets. To achieve this, the agreements were initially drafted to be legally binding. Individual firms would make commitments, as part of their sectoral agreement, which they would not be able to escape at any point in the future. Similarly, the government would not be able to revise the sectoral targets, nor, therefore, the targets for individual firms, during the period of the agreement, even if new scientific evidence showed that a pollutant was more harmful than previously believed. In the end, the government could not accept having its freedom to act curtailed in this way and the agreements were drafted as purely voluntary 'declarations of intent', from which firms are free to resign.

Before the government could enter into detailed discussions with companies in each sector, it first had to establish a reliable body of information on the levels of pollution and the prospects for reduction. For each sector, consultants developed an 'emissons profile' of the overall waste and emissons in a baseline year (usually 1985). A separate economic assessment predicted growth in each sector up to 2010, thus establishing a 'business as usual' emissions profile. Discussions between the government and industrial organizations representing each sector – allied with independent assessments of the reductions which could be achieved through widespread application of best available technology[6] – were the basis for agreement on each sector's contribution to the overall target.

Having agreed how overall targets, e.g. a 45% reduction in NOx emissions by industry, were to be distributed among the industrial sectors, these sectoral targets had to be allocated in some manner to the individual firms participating in each voluntary agreement. This process is described in more detail in the next section, outlining the voluntary agreement in the primary metals industry.

4.4. A Voluntary Agreement: the Primary Metals Industry

The 'declaration of intent', i.e. voluntary agreement (VA), on pollution reduction in the primary metals industry runs to 50 pages. It includes a legal preamble, list of participating companies, administrative structure and rules, requirements for company environmental plans and an integral environmental target plan (IETP) for the sector as a whole.

4.4.1. The Integrated Environmental Target Plan

The IETP sets targets for reductions of emissions into the air, water and soil and incorporates policy on soil clean-up, energy conservation, odours,

---

[6] Many of these assessments were conducted by the Dutch 'engineer bureaux'. These groups of industrial consultants are nominally independent, but the VNO points out that there is pressure on any consultant to provide results which please the client, in this case the government. The Dutch Office of Technology Assessment was also involved in these surveys.

noise and internal environmental management systems. It is drawn up for the years 1994/95 (four years ahead), 2000 and 2010. The agreement indicates which targets are subject to regional modifications, due to local conditions acknowledged in provincial and municipal environmental policy plans. (Global climate pollutants such as CO_2 are not subject to regional modifications.)

The targets in the IETP are demanding ones. They have to be translated into targets for participating companies, and the progress of each company and the sector's aggregate progress towards the sectoral targets, have to be tracked. A dedicated institutional framework has therefore been created to administer the agreement.

4.4.2. The Industry Consultation Committee

The primary administrative body for the agreement is the Industry Consultation Committee (ICC). This is a joint committee of government and industry representatives which is funded by the government. It reviews the environmental plans of each company, assesses progress towards the overall goals of the IETP, sets out how existing organizations should contribute to the IETP goals, sets up task forces, where necessary, on technical or policy issues and advises on modifications to the IETP in the light of current progress or improved projections. The ICC can take account of major external shifts in the business environment – such as the recession in the early 1990s – by modifying the sectoral targets. If an individual company has economic problems, the ICC will approach the Industry Support Commission for the NEPP, a newly created public body, for advice on adjusting that company's targets.

Industry is represented on the ICC by a newly created association – the Foundation for the Primary Metals Industry and the Environment. In other sectors suitable associations already exist, but the established metals industry association had much wider coverage than the relatively small number of firms concerned with primary metals production. One of the first tasks of the new association was to produce a 'model' company environmental plan for its members to follow in drawing up their individual plans.

4.4.3. Company Plans: Planning for Innovation

The majority of the primary metals manufacturers in the Netherlands have agreed to participate in the sectoral agreement. Among them are the Dutch branches of several foreign companies, such as Hunter Douglas Europe and ALCOA Nederland. An individually tailored company environmental plan, along the lines of the model plan developed by the industry association, sets out their long- and short-term commitments.

Each company plan identifies tasks intended to meet specific objectives, such as installation of equipment to reduce copper in water effluent by 30%, for the next four years, and projects activity over the four years beyond that, including any analyses required to set future tasks. The proposed tasks must allow the company to maintain 'reasonable profitability'. The company submits its draft plan to its usual licensing authorities, which must deliver a mutually agreed opinion within three months. The company then has two months to finalize the plan. If the licensing authorities feel that the plan does not go far enough, they can impose stricter requirements on the company. The company can appeal against this decision, on the grounds that its plan is consistent with the declaration of intent underpinning the sectoral agreement. The plan – excluding any commercially confidential parts – is sent to the ICC and is made public, as is the licensing authorities' report on the plan. This allows the ICC (and the public) to track the implementation of the overall sectoral plan. Each year, the company must report to the ICC on progress achieved so far. This includes an account of actual pollution in the last year, projected measures for the forthcoming year and the expected reduction in pollution as a result of these measures.

Measures to reduce pollution are classified into three groups:

- definite measures which can be easily implemented, i.e. well established technologies;
- conditional measures where one or more well-defined conditions relating to technical or economic feasibility must be met;
- uncertain measures which generally require some further analysis or research before a judgment can be made on their effectiveness.

This classification scheme explicitly links environmental objectives to innovation, but in a flexible manner which allows adjustment for the inevitable surprises of the innovation process. All manner of change can be accommodated, from good housekeeping measures to adoption of proven technologies, adaptation of technologies which have been proven in other settings, results of near- and long-term in-house research and results of government sponsored research.

4.5. Making the Agreements Work

4.5.1. The Licensing Regime

Ultimately, sectoral plans are implemented through the pre-existing site licensing procedures. Under the Environmental Protection Act, the licensing authority is assumed to be the municipal executive. The government can make an order to push this authority upwards to the provincial government or even the Environment Ministry, when regulation of certain pollutants or activities has to be coordinated at a regional or national level. In addition, the Pollution of Surface Waters Act gives licensing powers to regional water management boards. As a result, prior to the NEPP any one company might have dealt with several licensing bodies. This will generally remain the case for companies which are now signatories to an agreement. A company's environmental plan, accepted as part of the sectoral agreement, will have to be approved by each of the licensing authorities it normally deals with. The ICC has the task of setting out how the various authorities should cooperate, and ensuring that each understands the need for flexibility in the licence requirements for companies participating in the agreement.

The classification of pollution reduction measures within the agreements into definite, conditional and uncertain measures helps to structure the authorities' approach to licensing. Current licences are made contingent on companies applying 'definite' measures within the time scales they specify in their plans, and working to convert 'conditional' and 'uncertain' measures into definite ones. For their part, the companies benefit from a more consistent approach by the authorities, which are expected to refrain from imposing requirements that are not foreseen in the eight-year time horizon of the company plan.

4.5.2. Coping with Free-Riders

Ideally, all the firms constituting a particular industrial sector would be party to the sectoral VA. In practice, this has not proved to be the case. For example, by late 1993, only 35 out of approximately 50 firms comprising the Dutch metals manufacturing industry were members of the sectoral agreement. This raises the possibility that those firms which remain outside the agreement might escape any environmental obligations, i.e. they would be free-riders, benefiting from the efforts of others.

To avoid this, the government is working with the licensing authorities to put pressure on non-participating companies. Licensing authorities are expected to discriminate between firms which are members of the agreement and those which are not, in the course of their normal licensing procedures. Firms which have not joined an agreement in their industrial sector should be forced to meet the sectoral targets, for every pollutant and for waste volume, noise, etc. In contrast, the authorities should be allowing some flexibility to signatories to the agreements, on the grounds that they are part of a group of companies which will meet the same targets on average. As firms have realized the advantages of trading off better environmental performance by one measure against poorer performance by another, they have been increasingly willing to join the agreements.

4.5.3. Transitional Difficulties in the Licensing Regime

To date, one of the main weaknesses of the voluntary agreement system has been the reluctance of some licensing authorities to cooperate. Some local authorities feel that they were not adequately consulted on the national targets and on the VA process itself. Some are wary of local reactions if they treat a polluting firm leniently as a result of the trade-offs implicit in the VA approach. Others have shown an unwillingness to apply the full range of sectoral targets to firms which are not participating in an established agreement. The role of the licensing authorities in implementing the NEPP is not laid out explicitly in legislation, but rests instead on the procedures drawn up by the ICC of each agreement, and so the Environment Ministry has been concentrating on improving communications between the two groups.

Underlying these difficulties is the complexity of re-orienting a disparate licensing regime, which has evolved somewhat haphazardly, to the task of serving a comprehensive national strategy. Around the licensing regime a set of guiding principles and a body of accepted wisdom have evolved, and often these seem to be in conflict with the requirements of the NEPP approach. Traditionally, authorities have insisted on environmental performance which can be achieved with 'state-of-the-art' technologies, where these tended towards a rather conservative interpretation based on commonly applied technologies. Cost considerations have tended to figure large. In an attempt to correct this, and provide support for the more ambitious approach embodied in the NEPP, the revised Environmental Protection Act (1993) requires that licences are based on providing 'the greatest possible protection for the environment ... unless this cannot be reasonably achieved'. The Environment Ministry has coined the acronym 'ALARP' – As Low As Reasonably Possible – to describe what licensers should be aiming for. They hope that this will shift the balance of interpretation of state-of-the-art, or 'best available technology', in favour of environmental protection.

If the local authorities' ingrained licensing habits are to change, national dissemination of usable, up to date information on best practice is essential. To support a consistent approach to interpreting ALARP, the Environment Ministry has established an information centre at the Public Health and Environment Research Centre (RIVM). This is a source of reference standards which the government considers to be 'reasonably achievable', and information on associated technologies. Keeping the information up to date has been a major difficulty. The information centre has also been ignored by many local authorities since they generally dislike any kind of centralizing. In response, the Environment Ministry is considering the possibility of giving the reference standards legal status, so that the formal definition of ALARP would be consistent throughout the country.

4.6. Voluntary Agreements on Energy Use

The Dutch target for CO_2 emissions is stabilization at 1989/90 levels by 1994/95 and a reduction of 3–5% by 2000. The NEPP requires that the industry target group achieves a 20% reduction in energy intensity by 2000. In order to achieve this, the government decided to apply the voluntary agreement mechanism to reducing energy consumption in industry. By

November 1993, 17 'Long-term agreements on energy efficiency measures' had been set up with industry sectors such as refining, paper, textiles, chemicals and brewing. A further 17 agreements were under discussion. Where these sectors also have agreements on general pollution reduction, the energy-efficiency agreements have been included in the Integral Environmental Target Plan.

These energy efficiency agreements are backed up by the licensing regime in the same manner as the pollution agreements. To ensure this, Section 1.1(2) of the Environmental Protection Act (1993) establishes that consumption of energy and other resources is an environmental issue. As a result, licensers must apply the ALARP test to the energy performance of technologies and processes, when considering licence applications. Once again, the attraction for participating companies is the potential to trade off reductions between members of the agreement.

4.7. Innovation and the NEPP

4.7.1. Innovation Support Mechanisms

Dutch environmental policy explicitly recognizes the role of innovation in achieving the ambitious goals of the NEPP. Innovation is the key to the hoped-for progression from emission-oriented to volume- and structure-oriented abatement methods, without which the major advances being sought in environmental performance would be impossibly expensive. As such, it clearly forms part of the intellectual underpinning of the NEPP. It is no surprise, therefore, that a range of formal support measures for innovation have been created in recent years.

Tax relief on innovative technologies is available through the instrument for 'Accelerated depreciation on environmental investment'.[7] Over 300 approved pieces of equipment (including energy-saving technologies) form the 'environmental list'. If a company purchases a piece of equipment on the list, it can fully depreciate its investment for tax purposes in as little as one year. (Normally, the investment would be depreciated at a steady rate over the 'useful life' of the equipment). The 1993 list included a biological

---

[7] Ministry of Housing, Physical Planning and the Environment, *Accelerated Depreciation on Environmental Investment in the Netherlands*, VROM, June 1993.

sulphide oxidation reactor, cryocondenser for CFCs, thermal wood-preserving unit and rotating natural gas furnace.

One of the organizations responsible for running the accelerated depreciation scheme is NOVEM – the Netherlands Agency for Energy and the Environment. NOVEM ensures that prospective technologies for the list are innovative and of environmental benefit or energy saving. In theory, the existence of the list provides a strategic opportunity for these companies to expand their technology development activities. However, NOVEM report little evidence of this sort of behaviour. The technology suppliers tend to be small firms with a limited ability to formulate effective long-term strategies. This is also reflected in the role they play in the NEPP: waiting to be asked what technologies are on the horizon, rather than influencing target-setting processes from the outset. Their trade organization – the association of environmental equipment suppliers (VLM) – has very little political influence and is in a poor position to take on this role.

Making companies aware of environmental technologies and processes is an area which is also being tackled openly. The Netherlands has a network of 18 'innovation centres', providing information on innovative technologies, primarily to small companies. These are now putting much of their effort into promoting environmental technologies. Since 1988, a joint industry/government 'National Centre for Cleaner Production' has been supporting environmental managers in firms by disseminating information and providing courses on waste prevention, environmental management, etc. Finally, a joint industry/government network of environmental service centres (BMDs) has been created to assist member firms with the introduction of environmental management systems.

4.7.2. Implicit Impacts on Innovation

Far more important than the explicit policies to support innovative technologies are the impacts on innovation implicit in the whole NEPP approach. The NEPP has the potential to bring about a deep cultural shift in how industry, and the nation as a whole, view environmental policies. This is causing firms to focus their innovative resources on the goal of sustainable development.

Three features of the NEPP are changing industry's attitudes towards the environment. First, there is the impressive breadth of support for the overall approach. There has been cross-party parliamentary support for the NEPP strategy and wide public consultation and acceptance, all of which lends strength and credibility to the policy. This has created an air of inevitability which persuades most firms that cooperation is preferable to resistance. Many firms, and the VNO, took this view early in the process of formulating the NEPP.

The second feature is the long time horizon of the NEPP. In 1990, targets were set for 20 years ahead, and these have cascaded down through sectoral agreements to individual company environmental plans. The credibility of the overall NEPP strategy lends credibility to the targets. All businesses crave a stable environment and in the past the suddenness of much environmental legislation has generated much of the resistance among firms. The new approach provides a stable framework over the long term. Firms begin to regard environmental issues as less an uncontrollable business risk than a potential opportunity, and are better able to plan relevant investments. Third-party suppliers have the assurance to develop improved technologies over much longer periods than was previously commercially sensible.

Finally, the process of implementing the NEPP is explicitly based on cooperation between industry and regulators and on achieving flexibility through voluntary agreements. (Once again, this feature is supported by the credibility of the NEPP, since firms are persuaded that sanctions against non-participating firms are likely to have widespread support.) Firms feel they have a greater choice over the means by which they respond to environmental pressures and their attitude to these becomes less hostile. Additionally, as the two sides cooperate, a more open flow of information on costs and technologies evolves. The iterative annual rounds of forecasting and reviewing the company plans and identifying 'definite', 'conditional' and uncertain measures, formalize a 'technology foresight' approach, which allows innovation to be planned and accounted for.

Figure 4.1 shows the interaction between the overall policy context, the choice of voluntary agreements as the framework for regulation, and innovation. Broad support within society provides the credibility which underpins the long-term objectives and voluntary agreements. Increased freedom of choice and stability promotes a change of mindset within industry. Industry's increased support for environmental goals, and the NEPP strategy

Figure 4.1 Interaction of Policy, Regulatory Instruments and Innovation in the Dutch NEPP

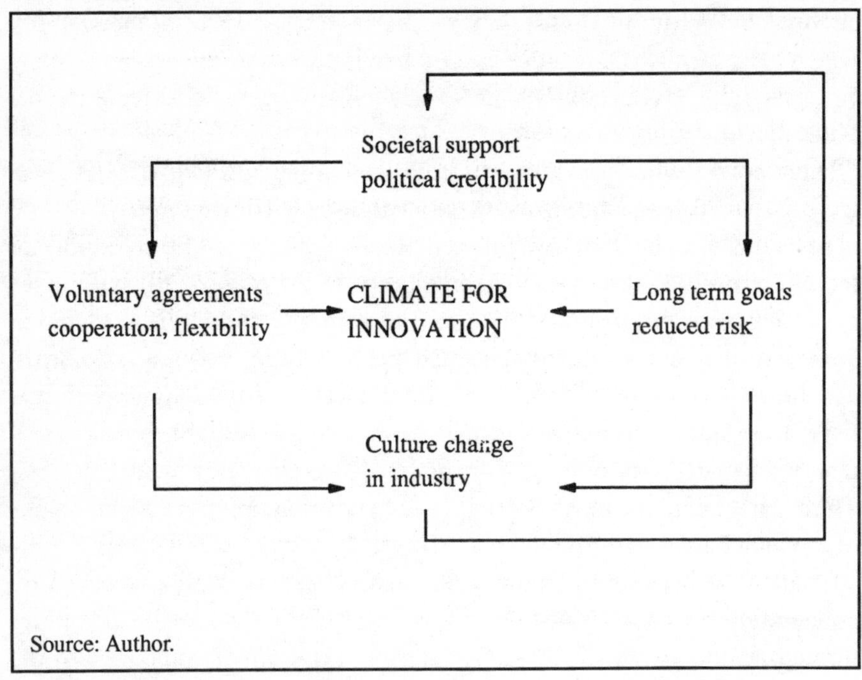

Source: Author.

in particular, adds to the degree of support within society, further enhancing the credibility of the policy. An improved climate for innovation arises almost as a fortuitous by-product of a society-wide shift in attitude and approaches to the environment.

4.8. Early Difficulties with the NEPP

This analysis allows us to identify where the Dutch government has made some mistakes in implementing the NEPP. Early in the process of formulating the plan, a delicate balance was struck between the government's environmental goals and the advantages to industry of the new, friendlier approach. This consensus was severely disrupted by a change of government in 1989, shortly after publication of the NEPP proposal. The new government was, as before, led by Ruud Lubbers and his Christian Democrats, but the new junior coalition partner was the Labour Party. The previous coalition partner, the Liberal Party, had objected to reduced subsidies to car-owners which would result from the NEPP, and then withdrew its support for the government and forced an election in which it lost ground.

In keeping with the wishes of the Labour Party, and the apparent mood of the electorate, the government announced and passed an update to the NEPP – the NEPP Plus. This contained several new, national targets which had not been agreed with industry. Chief among these were a CO_2 reduction target, at a time when industry was ferociously resisting a unilateral Dutch CO_2 tax, and a shortening of the time scale for reducing acid emissions to the 'sustainable' level. In response, the VNO withdrew its formal support for the NEPP Plus, denting the credibility of the new strategy.

According to Figure 4.1, this setback can be seen as a disruption to the process of creating stability in environmental policy – the right-hand route to a change of attitude within industry. Offsetting this interference with the long-term goals was solid progress on the form and operation of the voluntary agreements – the left-hand route in the figure. By 1993, most of the voluntary agreements originally hoped for had been signed, even where these incorporated targets based on the disputed, revised goals. So far, the twin approach has proved to be a robust one.

For the future, interference with the long-term national goals by politicians seeking to curry favour with narrow interest groups represents the biggest risk to the NEPP. Genuine credibility on this issue can only be created over time, unless a future government is willing to enter into binding contractual agreements with firms. The voluntary agreement mechanisms are less susceptible to interference since they are backed up by a detailed, painstaking process. The only foreseeable danger here is that licensing authorities will neutralize the sanctions which encourage firms to participate, by refusing to apply the maximum sectoral standards to non-participating firms.

Loosening of standards can be as damaging to the consensus supporting the NEPP as raising standards without consultation. This was clearly demonstrated in the case of standards for soil contamination. Prior to the NEPP, a major soil protection plan was developed to ensure that any material finding its way onto Dutch land would meet the strict limits for contamination by heavy metals and chemicals which apply to agricultural soil. This requirement applied to waste aggregates, such as gravel (including rubble from demolished buildings), slag and fly ash. The plan required that these should either be reused, in buildings, etc., or should meet the same standards as agricultural land, even if they were being used as fill material in dykes or roads. The reasoning was that with sufficient time the material would eventually diffuse into the surrounding land.

Inconsistency over one particular standard caused a great deal of innovative effort to be wasted. Polycyclic Aromatic Hydrocarbons (PAHs) are by-products of combustion and can be found in the brickwork of old Dutch houses where coal was the traditional fuel. The soil protection standard announced in 1989 implied that all waste material from demolished buildings must have a PAH content of less than 10 ppb. Companies specializing in soil contamination developed new methods for achieving this standard, such as thermal bioremediation. However, during negotiations on implementing the NEPP, the government decided that in the case of PAHs the standard applied to agricultural land was an inappropriate benchmark. They increased the maximum acceptable concentration for waste aggregates to 75 ppb, with the result that there was no longer a market for the new technologies which were developed to meet the tighter standard. This shook the faith of many companies involved with soil pollution, a topic which forms a substantial part of the NEPP.

These difficulties illustrate how important it is for governments to take a consistent approach when introducing a policy, such as the NEPP, which is both far-ranging and dependent on a widespread consensus within society. Some comfort can be taken from the fact that both of the problems described above arose when the NEPP was in its formative stages. The two were in a sense a result of a collision between existing policies and the new NEPP approach, where the common problem was one of integration. Similar difficulties will be faced by any country wishing to follow the Dutch example, but each attempt will increase the body of experience – and the chances of success.

4.9. Summary

This study of the Netherlands is above all the story of a bold attempt to create a sustainable economy. Before committing themselves to this course of action, policy-makers made a dispassionate appraisal of the effort required. They are well aware of the enormity of the task involved in radically transforming the functioning of their society and have concluded that innovation holds the key to long-term success.

Industry has taken responsibility for managing its own transformation: the system of voluntary agreements provides the flexibility for firms to set their own goals and to use their own resources to innovate and effect change in the manner which suits them best. The promise of this flexibility was crucial in securing industry's cooperation in setting realistic national goals for the NEPP.

Close involvement in formulating the NEPP also extended to environmental groups, citizens' groups, and local and national authorities and political groups. In this way, the NEPP is attempting to create a broad social consensus of the kind which has evolved more slowly in Denmark. This consensus underpins the credibility of the NEPP goals. The long time horizon of these goals reduces the commercial uncertainty and risks associated with environmental policy and so helps to secure industry's cooperation with the voluntary agreements. So far, the mutually reinforcing relationship between policy and mechanisms in the NEPP has survived several major challenges.

Some observers have criticized the NEPP on the grounds that it has taken longer to put into practice than was originally hoped. In my view this is inevitable. The NEPP represents the world's first serious effort to create a sustainable industrialized economy. It needed a massive effort to gather data on the Dutch environment and industrial structure. It necessitated new thinking about regulatory mechanisms, associated institutions and the relationship of these to existing mechanisms and institutions. It required the creation of a national consensus that sustainable development should be a major social objective. None of these efforts is complete, nor could they reasonably be expected to be completed in less than a decade. Several governments will need to come and go, each lending its support to the NEPP while avoiding excessive tinkering with its details, before industry is finally convinced that it will endure.

Chapter 5

Germany

5.1. Introduction

Environmental policy in what was the Federal Republic of Germany dates back to the late nineteenth century. Until the Second World War, environmental policy was known by the term 'air and water hygiene'. Particularly in the industrial region around the Ruhr valley, air pollution was a subject of public interest as early as the 1920s (for example, because of its impact on local and regional forests), and measures were demanded to protect the environment. The pressing problem which arose in this region led in the 1950s to clean air protection initiatives by the legislature in North Rhine Westphalia, where the Ruhr valley lies. The legislation which was developed in this state in the 1960s, and the related programme of environmental research, were the basis for federal legislation at the beginning of the 1970s, when they became an integral part of the Federal Clean Air Act.

Up to the beginning of the 1970s, environmental policy was organized on a regional basis in the Federal Republic, being chiefly the responsibility of the individual *Länder* (states). The subject received a decisive impetus at the end of the 1960s, when public activism on environmental pollution and environmental protection spread from the United States, and a special environmental protection division was set up at the Federal Ministry of the Interior. This was also the first step towards the establishment of a separate environmental policy area.

The pace of development of environmental policy intensified in the early 1970s, through the adoption of the first federal environmental programme and the establishment of the Council of Environmental Advisors (SRU), whose mandate was to report on the state of the environment and pinpoint undesirable developments and ways of avoiding them. In 1974, the Federal Environment Agency (UBA) was established in Berlin. Reporting to the

Federal Ministry of the Interior, UBA was given a mandate to support the minister responsible for environmental issues in all tasks involving environmental policy, and to develop the scientific expertise which enabled it to carry out this advisory function. A separate environment ministry was not set up at Federal level until 1986, and then partly as a reaction to the Chernobyl nuclear accident in the Soviet Union. The Federal Ministry for the Environment, Nature Protection and Reactor Safety (BMU) is based in Bonn with the rest of the government machinery. The Environment Agency is now attached to the Environment Ministry.

The Council of Environmental Advisors has tracked sources of information on the public perception of environmental issues, such as national opinion polls, since the 1970s. In its 1978 report, the SRU noted that environmental protection was by then a well established and highly rated political priority of the population. This continued to be the case throughout the 1980s. The SRU stated in its 1987 report that, together with overcoming unemployment, environmental protection was listed as a priority in the opinion polls. The advisors believed that the environment was not just a temporary concern, but rather had taken up a permanent place on the political agenda. The general public's awareness of environmental problems was reflected in the political landscape, with the Green Party capturing a significant share of votes in national and state elections, especially in the mid-1980s.

In the 1970s, the major problem faced by regulators was very high local and regional air pollution and many areas of highly contaminated water. Measures to reduce these impacts often had immediate beneficial results, but, on closer analysis, proved to have simply shifted the problems elsewhere. One example of an environmental policy which had a very effective local impact but led to 'problem shifting', was the clean air policy during the 1970s. A decision to raise the height of chimneys on coal-burning power stations and industrial plants was one of the main elements of a strategy which produced considerable improvements in local air quality. However, the widespread dispersion of airborne pollutants affected areas previously free of environmental damage. In particular, this strategy is widely held to be responsible for much of the forest damage caused by acid deposition.

In the first half of the 1980s, forest decline (*Waldsterben*) was the most important environmental topic and, in view of the high public interest, an intensive political effort was made to get this problem under control. Largely through a rapid installation of end-of-pipe pollution control technologies in fossil-fuelled power plants, a considerable reduction in sulphur dioxide pollution was achieved in the 1980s. After research into forest damage had supplied the first indications that NOx also contributed to large-scale forest damage, preventive action was extended to vehicles. This began in the mid-1980s with measures to reduce NOx emissions from car exhausts.

Later in the decade, the new priority areas were identified as environmental problems caused by agriculture, identification of contaminated sites requiring clean-up, waste disposal and global problems such as the depletion of the ozone layer and the greenhouse effect. Traffic-induced emissions grew in importance, as measures initiated to reduce pollution were offset by the increased motor vehicle traffic, and the resulting high summer ozone levels became a new subject of concern. Environmental pollution in the former German Democratic Republic – which had reached catastrophic levels in places – represented a new challenge for German environmental policy.

Trends in the energy sector in the 1970s and 1980s had major spillover effects on environmental policy. The dominant issues in the sector were the two oil crises and the nuclear energy controversy which had been smouldering since the beginning of the 1970s. The efforts to achieve a more rational use of energy spurred by the two oil crises have led to considerable environmental relief, prevented additional environmental pollution and pushed the problem of conserving resources into public awareness. The environmental movement was very strongly associated with the anti-nuclear movement, as both had come to the fore – along with issues such as disarmament, women's rights and North–South relations – as part of the new, radical political agenda of the 'sixty-eighters', who wrought more permanent cultural and political changes in Germany than in most other industrialized countries.

By the end of the 1980s the Environment Ministry had concluded that the reactive environmental policy of 'closing the stable door after the horse has bolted' was nearing the limits of its effectiveness. Pursuing an environmental policy focused on single issues had led to problem shifting, as with the power station example or that of sewage sludge generated by waste water purification. In the ministry's view, an environmental policy founded primarily on end-of-pipe measures had led to a temporary stabilization of the environment, but ministry officials expected continued economic growth to cause further environmental deterioration in the medium or long term.

The failure of reactive policies based on narrow definitions of the problems forced the Environment Ministry to conclude that, for the future, a cross-sectoral, preventive environmental policy, oriented to the criteria of sustainable development would be essential. The aim is to ensure conservation of raw materials, energy and environmental resources, so as to eliminate any detrimental effects of today's society on the living conditions of future generations. Simultaneously, this requires an intensive interlinking of environmental policy, economic and structural policy, and raw materials, energy and transport policies. This chapter explores the first steps taken in this direction, and the role of German industry in this process.

5.2. Structure of Environmental Policy

Germany is a federally organized state in which the legislative roles are distributed between the federal government, the *Länder* and the local authorities. At the national level there are two main legislative bodies: the Bundestag, composed of directly elected MPs, and the Bundesrat, made up of representatives from the *Länder*. The various responsibilities for environmental legislation are distributed in the following way:

- the *federal government* has the responsibility for waste management, air pollution, federal waterways and industrial pollution;
- the federal government makes framework legislation and the *Länder* provide detailed regulations for nature conservation, landscape management and protection against water pollution;
- the *local authorities* have the responsibility for the local environment, in as far as it is affected by urban development planning (regulation of building activities, traffic planning, cleaning up contaminated soils, etc.).

The *Länder* and the local authorities are responsible for enforcing and applying all environmental legislation, for example through licensing of industrial sites, with the exception of nuclear safety, which is the responsibility of federal authorities. To be successful, this structure demands effective, continuous cooperation between the federal government and the *Länder*. As in other policy areas, cooperation between these two levels of government centres on the Bundesrat.

Ministers and their staff propose legislation for approval by the Bundesrat and Bundestag. Those with an important influence on environmental policy are the Environment Ministry (BMU) and both its Environment Agency (UBA) and the Federal Office for Nature Conservation (BfN); the Economics Ministry (BMW), responsible for industry including the energy industry; and finally – holding the purse strings – the Finance Ministry.

Since 1982, a series of coalitions dominated by the Christian Democratic Party (CDU) have formed the federal government. Ministerial posts are shared among the parties making up the coalition and, as a result, ministries are often used as platforms for airing party policies and drawing distinctions between the members of the coalition. If a particular issue has a high public profile, several ministries may produce policy documents which set out radically different positions. Generally, these are reconciled in the chancery – the office of the chancellor – and in parliament, but the subsequent, 'consolidated', compromise legislation frequently receives less publicity than the earlier statements. Outside Germany, observers often mistake this ministerial kite-flying for official government policy.

One consequence of this partisan approach to policy-making is that certain ministries are habitually reluctant to coordinate and consult with one another on legislative proposals. Environment Ministry officials confirm that they have avoided consultation with other ministries on environmental legislation for which they have been responsible. For a long time the lack of interministerial coordination was exacerbated by the personal political strength of the former Environment Minister, Klaus Töpfer. Töpfer was an enthusiastic and vocal defender of the environment. Other ministries have in the past felt that they were in a poor position to argue against Töpfer's initiatives because of the public strength of feeling on the environment – demonstrated by the size of the electoral support for the Green Party. Töpfer's high profile stance on the environment helped to make him one of the best known and most popular of Germany's politicians.

The Environment Ministry's autonomy was bolstered in the 1980s by the Bundesrat's enthusiasm for environmental legislation. Many deputies from the *Länder* seemed to be under the impression that the environment was invariably a vote-winner. This additional source of political support added to the feeling within the Environment Ministry that it would be possible to reduce the involvement of other ministries in the policy-making process to a bare minimum, and avoid any consultation with industry representatives.

5.3. The Licensing Approach – Best Conservative Technology?

In the 1970s and 1980s the Federal Republic relied predominantly on regulatory instruments such as the establishment of emission standards. Each of the *Länder* has its own environmental inspectorate, to carry out its role of enforcing the federally mandated standards.

Germany's regulations on air pollution provide a good example of its generally comprehensive and detailed approach and have arguably had a greater impact on industry than other areas of environmental regulation. The Federal Clean Air Act of 1974 (amended in 1985) gives authority to the *Länder* and local authorities for protection of the environment and public health and avoidance of pollution. Detailed measures are to be based on 'state-of-the-art', i.e. best available technologies (BAT). The law does not make any reference to economic costs.

The Environment Ministry is charged with putting the flesh on the bones of the law, through federal ordinances. The most important of these is the Administrative Regulation pertaining to Technical Instructions on Air Quality Control', or 'TA Luft', of 1986. This sets out ambient air quality and deposition standards, referred to as '*i*mission' standards, and source-specific *e*mission standards for particulates, organic and inorganic gases and carcinogens. The air quality standards are based on technical advice from the Environment Agency in Berlin. Emission standards are determined largely by an evaluation of current BAT, based on information supplied by both the Environment Agency and equipment suppliers. Abatement costs are, strictly speaking, not taken into account, but the Environment Ministry nonetheless makes some attempt to estimate the marginal costs of abatement for some pollutants, and takes this into account when setting final standards. New facilities are required to utilize BAT, even in areas where ambient pollutant levels are well below the maximum permissible limits.

In practice, the TA Luft has proved to be rather inflexible and has been inconsistently interpreted and applied by licensing authorities. Many of the difficulties stem from the long delays between major updates: the TA Luft replaced a similar ordinance which had been in use for ten years, and the Environment Agency expects the TA Luft to be in use for a similar length of time.[1] At the regional level, the inspectorates within each *Land* become very familiar with each vintage of regulations and take a long time to learn the new regulations when a major update such as the 1986 TA Luft is announced. Inspectorates are in general fairly conservative, a situation which may be exacerbated by political links between some large industrial firms and certain *Länder*.

In one sense, however, the licensing process does exert a more continuous pressure on authorities to revise their working definitions of BAT. In areas with pollutant levels approaching the official air quality limits, a new process or facility might be forced to install equipment which achieves emission levels well below those indicated in the TA Luft, in order to gain a licence. If this 'advance' becomes widely known, firms wishing to establish similar processes elsewhere may come under public pressure to achieve this new standard, or face a public inquiry. In such circumstances, licensing authorities have been known to insist on the tighter standard in order to avoid any political fallout, although strictly speaking the authority should only depart from the TA Luft standard when there is a local air quality problem. Similar processes can be observed in other areas, for example, the interaction between water quality standards and technologies.

There is, then, some ratcheting of standards of the kind which has effectively tightened standards in Japan, but it is different in nature.[2] In Germany, a severe, local problem may require secondary and even tertiary clean-up measures which are then demanded at other sites for political reasons, whereas in Japan 'windfall' gains from equipment which out-performs design criteria

---

[1] Inevitably, given its long life span, some of the TA Luft's provisions have been amended: e.g. the emission standard for dioxins from municipal incinerators was updated in a government ordinance of April 1990.

[2] Chapter 10 describes how SO_2 emissions from German power stations have been drastically reduced, not as a result of licensing processes, but because of the liabilities which utilities force upon their design and engineering contractors.

are quickly assimilated into the next round of licensing, at little additional cost to industry. Another significant difference is that whereas authorities in Japan continuously exchange detailed information on licensing activities, there is – in the view of the German Environment Ministry – relatively little of this kind of communication between the German *Länder*.

Not surprisingly, given the rather fragmented and conservative approach to licensing, Environment Ministry officials are not aware of pollution equipment manufacturers pursuing a deliberate strategy of influencing the regulatory process in the hope of generating markets for their products. However, this may change as the environmental industry becomes more concentrated. Over the last several years large industrial firms, particularly electric power utilities such as RWE – Germany's largest – have been acquiring numerous small environmental firms. This is creating a group of much larger firms in the market with greatly increased political weight, and the management capacity to become involved in lobbying. This may begin to provide a counterweight to more conservative industrial forces.

At the federal level, the lack of dialogue between the Environment Ministry and industry has been an impediment to the flow of information on new or improved ways of achieving environmental goals. Some of the political reasons for this poor relationship were outlined in the previous section. In addition to this, it seems that the ministry has in the past been content to turn to its Environment Agency for most of its advice on technological possibilities for meeting its environmental objectives. The wisdom of doing so is questionable. The agency existed as a separate institution for many years prior to the formation of the Environment Ministry and has perhaps not fully adjusted to its current role. This is compounded by the fact that the agency, in Berlin, is geographically remote, and was even more isolated prior to reunification. This has meant that the flow of information between the ministry and the agency has often been less than ideal. Its geographical isolation also limited the extent to which the Environment Agency could form useful relationships with West German industry.

5.4. The Costs of Polarization

Expenditure on environmental protection by industry almost doubled between 1980 (7.8bn DM) and 1987 (14.1bn DM).[3] The BDI (the influential association of large firms) estimates that combined private and public sector expenditure amounted to 1.74%of GDP in Germany in 1991, compared to 1.36% in the United States and 1.02% in Japan. The Environment Agency was becoming aware of these escalating costs and anticipated that costs to industry of complying with regulations would accelerate. The agency advised the ministry that, in many areas which had been subject to regulation for a long time, the existing regulatory approach would only gain further improvements at greatly increased cost: the marginal costs of abatement were rising steeply. At the same time the Environment Ministry was taking the decision to move away from its reactive policies which were traditionally concentrated on single issues, towards a new, comprehensive strategy for pollution control, based on sustainable development and cross-sectoral, preventive policies.

As the government began to think about how best to implement this new environmental strategy, Germany started to slide into the recession of the early 1990s, precipitated in part by reunification with East Germany in 1990. Industry's complaints about the costs of environmental regulations became more vocal, and the Environment Ministry was compelled to take notice. Industrial organizations, particularly the BDI, called for a moratorium on new environmental regulations, with some success.

However, others saw an opportunity for exploring new policy-making processes. The Environment Ministry sought to establish a greater dialogue with industry and received a positive reaction from the Association of German Chambers of Commerce (DIHT), one of the main representative bodies for small firms, although the BDI showed less interest in creating a closer relationship. The DIHT argued strongly for a new approach to environmental issues, based on voluntary actions by industry. This approach was adopted on a trial basis as the first stage of the plan for a 'closed-cycles economy'.

The DIHT believes that the Environment Ministry is now much less confrontational and that much of the strong rhetoric of the recent past, evident

---

[3] In real terms, at 1980 prices. Source, Federal Statistical Office.

on both sides, has disappeared. Even so, the relationship between the two is 'not good'. The DIHT claims that it does not seek to resist environmental goals, and that industry learned a long time ago that it 'cannot beat public opinion and the government' – however, the DIHT does want a new strategy and new kinds of regulatory instruments.

5.5. Cutting the Wasteful Economy

The 'closed-cycles economy' plan was Klaus Töpfer's strategy for a cross-sectoral policy, focused on resource conservation and waste reduction, consistent with the guiding principle of sustainable development. The concentration of efforts on waste reduction and conservation of resources stems partly from the Environment Ministry's observations of trends in pollutants.

The Environment Ministry was struck by the fact that a decoupling of economic growth and certain types of pollution took place in West Germany in the 1970s and 1980s. This applies in particular to some of the classic air pollutants. While real GNP rose by over 50% between 1970 and 1990, sulphur dioxide emissions were reduced by 75%, dust emissions by 65% and carbon monoxide emissions by 43%. Whereas the reduction in dust emissions was mainly achieved in the 1970s, significant reductions in sulphur dioxide and carbon monoxide emissions took place in the 1980s.[4] The ministry expects to see further reductions in these pollutants as a result of existing regulations. These reductions were achieved through more rational use of energy (decoupling of energy consumption and growth), the intensified use of low-emission fuels (natural gas, nuclear energy) and the enforcement of air pollution abatement measures in the first half of the 1980s. Although NOx emissions had risen by 10% over two decades, the recent trend was downwards, from a peak of 27% above 1970 levels in the mid-1980s. Existing vehicle regulations were expected to reduce further overall NOx emissions, over the short term at least.

However, there was no such decoupling of economic growth from the volume of wastes being generated.[5] Industry continued to produce

[4] 1990 Report on the Environment, BMU, Bonn, May 1990, cited in *Science Responds to Environmental Threats: Germany Country Study*, OECD, Paris, 1992

[5] Ibid.

considerable amounts of waste in the form of products and packaging which become domestic waste. Bottlenecks in waste management, caused, for example, by lack of landfill sites and public opposition to new incineration plants, forced the government to pursue a more aggressive waste prevention policy.

The resulting waste framework law was passed by the Bundesrat in 1994, after lengthy discussion and substantial amendments, partly to take account of negotiations on European Union directives on waste. The framework law should come into force by 1996 at the latest. It establishes the following hierarchy (in order of preference) for the disposal of materials used in products:

1. avoid waste, i.e. re-use the material;
2. recycle or burn to produce energy;
3. dispose to landfill, after incineration to reduce the volume of waste.

For certain products, e.g. passenger cars or personal computers, subsequent decrees will identify in detail what percentages should be re-used, recycled or disposed of.

As a prelude to this grand strategy, the government adopted the Ordinance on the Avoidance of Packaging Waste, in 1991. This makes manufacturers and distributors responsible for collecting, processing and recycling any packaging or containers sold with their product. In January 1993, when the ordinance entered into force, 60% of glass waste had to be collected and, of this, 70% was to be sorted, giving an overall recycling rate of 42%. Paper and card had an 18% recycling requirement. In July 1995, higher collection and sorting requirements will result in recycling requirements of 72% for glass, steel, aluminium, paper and card and 64% for plastic (up from only 9% in 1993).

In the interim period between the adoption of the ordinance in 1991 and its enforcement in 1993, the DIHT helped to organize a voluntary scheme which would enable approximately 600 of its member firms to enter into compliance with the ordinance. This scheme is run by a non-profit company, the Duales System Deutschland, or DSD, and began operation in 1991. The member firms – manufacturers and distributors – pay a fee to DSD and in return earn the right to print a green dot on all the packaging they use. The DSD collects and sorts household waste carrying the green dot (householders having played their part and disposed of these items in special bins), and arranges for it to be reprocessed.

Several major difficulties with the system have arisen since it began operation. Some are due to commercial blunders, others are perhaps more fundamental. First, collection and sorting costs are high, partly because there is no pressure on the DSD to control these costs, which are passed on to member firms, and ultimately to consumers. The DSD estimated that it would collect and process 5.6 million tonnes in 1994 at a cost of around 4 billion DM. This is double the normal cost of collecting household waste in Germany. Commercial inexperience led to near-bankruptcy with debts of around 1 billion DM by the end of 1993. Some of this was due to non-member firms using the green dot without paying the fee, and the DSD's failure to take any effective action against this fraudulent activity. However, the view of the Ministry of Economics was that much of the indebtedness was simply due to the DSD failing to send invoices to member firms.

These mistakes are redeemable. Far more serious are the doubts over the basic principles of the German government's approach. First, almost no consideration was given to recycling capacity. In 1993, for example, the DSD collected 360,000 tonnes of plastic packaging (the government had anticipated 100,000 tons), exceeding recycling capacity by 115,000 tonnes.[6] Much of this was dumped on recycled material markets in other countries, causing prices to drop and jeopardizing their market-led recycling efforts. Not only is the recycling capacity insufficient, but foreseeable demand for most recycled materials is far lower than the targets set by the government. Many of the materials now being recovered and reprocessed are more expensive than virgin raw materials. Influential commentators such as Professor Ernst von Weizsacker, President of the Wuppertal Institute for Climate, Energy and Environment, argue that the sensible approach is to raise raw material prices, through taxes, thus creating a bigger market for recycled materials.

If there is an innocent victim in this story it is the fledgling attempt by industry to demonstrate that it can voluntarily take responsibility for meeting environmental objectives. When the recycling scheme began, the DIHT hoped

[6] 'Germany's Green Dot Program Rebuked', *Recycling Times*, 15 June 1993, p. 16.

it would prove that voluntary action by industry can be effective. Many have taken the commercial problems the scheme encountered as evidence that, in general, voluntary measures will be ineffective. Even the Ministry of Economics, which one might imagine to be sympathetic to industry, is critical of industry's performance in this instance.

5.6. A New Industry–Environment Alliance?

The initial difficulties of the DSD scheme must be set against the wider background of industry–regulator relationships. As described earlier, the DIHT is aware of an improved mood of cooperation between industry and the Environment Ministry, with the potential to reconcile the polarized positions of earlier years (although this effort may be undermined by the tendency of some parts of industry, especially the BDI, to cling to the adversarial approach).

For its part, the Environment Ministry now talks of the need for a mix of regulatory instruments in which persuasion, raising consumer awareness and economic instruments are equally as important as legislated standards applied to industry, although these will still have their place. However, during the worst of the recession of the early 1990s economic reality forced the Environment Ministry to place an informal moratorium on any new ordinances. The ministry now talks of the need (in better times) to use all available regulatory options flexibly, i.e. as and when appropriate. Arguing about whether economic instruments are better than ordinances based on emission standards is, in the words of one official, 'the wrong debate'.

Some alternative regulatory instruments have a lengthy history in Germany. Eco-labelling, under the guise of the 'blue angel' award, has been employed since 1977. The Environment Ministry hopes increasingly to influence purchasing behaviour, through persuading manufacturers to educate the public about environmental issues in their advertising. Taxes of any kind are politically almost impossible to introduce during a recession, but the Ministry of Economics supports their wider use in the future, and industries have in the past been supportive, when taxes have suited their international ambitions.[7]

[7] See, for example, the discussion on differential vehicle taxes to encourage the uptake of catalytic converters, in Chapter 9.

How, then, does this improved mood of cooperation affect the draft product ordinances which are intended to implement the framework waste law? In contrast with the packaging ordinance, there is an opportunity with the product ordinances for industry and the regulators to make common cause. This arises because industry believes that the product ordinances may represent a powerful trade weapon. To take the example of the proposed ordinance on vehicles, any car sold in Germany will have to meet certain recycling targets or, in the case of a foreign manufacturer, be removed from Germany. This might be very costly, and in addition environmental groups will undoubtedly criticize foreign manufacturers who take back, and scrap, their cars and will try to persuade consumers to avoid them. The DIHT is convinced that a foreign manufacturer 'such as Rover' will ultimately be forced to adopt design features and recycling arrangements comparable to those already adopted by German firms.[8] However, German manufacturers will have been working on low-waste design issues for some time, and will have a significant advantage in this area. They believe that similar advantages for German competitiveness will be seen in the other industries affected by product ordinances, such as computers. Indeed, the DIHT comments that German industry 'deserves' such a windfall, as compensation for the investments it has had to make in environmental protection in the past.

It is possible that the individual decrees setting product take-back requirements could be implemented in the prescriptive form which was proposed early in the legislative process. After all, the industries involved will expect to see the costs of recycling systems offset by the loss in market share of foreign competitors, and so may not object too loudly. (In the personal computer market low-cost manufacturers based in the Far East may be particularly vulnerable.) However, widespread use of voluntary action might conceivably suit industry's purposes rather better. A period of voluntary recycling initiatives, with public funding for associated research and development activities, would allow firms time to design products which use less materials and are more easily recyclable. After a suitable time, mandatory standards which are tougher than currently envisaged could be

[8] Ironically, since this comment was made Rover has been bought by Germany's BMW and will presumably meet the recycling standards of the parent company.

swiftly announced causing perhaps few problems for German manufacturers, but seriously disadvantaging foreign competitors. In time, pressure could be applied to have the standards adopted by the European Union (in an extension of the EU's proposed waste packaging directives), once Germany had demonstrated that they were attainable.

An indication of the way things may develop was given in the middle of 1994, when the German automobile industry proposed a voluntary used car recovery scheme. The proposal was for the manufacturers to set up a collection and recovery scheme which would charge around DM200 to dispose of a vehicle and recycle as much of it as the operators considered practicable. The Economics Minister welcomed the scheme but Environment Ministry officials objected that it does not include detailed recycling commitments and requires legislative changes to vehicle registration procedures which will delay its introduction. Clearly the Environment Ministry remains mistrustful of voluntary agreements, but since its former minister repeatedly threatened to force through an automobile recycling decree and yet was never in a strong enough position to do so, it may have no option but to accept the industry's lead.

5.7. Conclusions

In common with the Netherlands, Germany is trying to establish the long-term goal of creating a sustainable economy as the heart of its environmental policy. However, relatively little systematic thought has been devoted to identifying the best mechanisms for achieving the ambitious goals of the framework waste legislation on which German efforts are currently based.

A genuine shift towards a sustainable economy implies fundamental changes throughout the economy and particularly in the industrial structure. In those other countries where this is a serious goal, close cooperation and involvement with industry at every stage is seen as a crucial factor for success. In the Netherlands, a complex, multi-layered system of agreements has been created and the national goals were determined through a lengthy consultation process. Japan is in the process of creating a national framework for sustainable development. This will be overlaid on a long tradition of flexible agreements between local industries and regulators. In both of these cases, innovation is explicitly recognized as an important means of meeting the environmental objectives with the minimum pain.

Unfortunately, German environmental policy-makers have a history of relatively poor relationships with industry, characterized by polarized, opposing views and mutual distrust. In this respect Germany is similar to the United States. (If Germany's environmental laws have been less absurd than those of the United States, it is because the expertise and technical understanding of the regulators in the German Environment Ministry and Agency far outstrips the knowledge of the legislators in the US Congress.) There is a danger that, with inadequate input from industry and inflexible regulations, the German economy will simply block up its existing waste outputs without giving industry a chance to learn how to develop the new processes and designs which will reduce throughputs of materials.

Chapter 6

France

6.1. Introduction

For much of the postwar period, the over-riding political concern in France has been to ensure its independence in the modern world and so, according to some observers, to establish French power and prestige abroad.[1] For French leaders, economic power has been a precondition for global influence, and industry is at the core of that economic power. Industry therefore has a political significance which goes beyond its usual role of securing a better standard of living for the population.

When Charles de Gaulle became president for the second time in 1958, he embarked on a programme of correcting the political and economic weaknesses which had dogged the Fourth Republic, established after the liberation of France at the end of the Second World War. De Gaulle reduced the power of the army, which had become a 'state within a state' through the pursuit of a succession of post-1945 colonial wars, and strengthened the power of the presidency and the government relative to the parliament, which had been dominated by self-interested political elites, or *notables*. A new constitution, overwhelmingly approved by a referendum in 1958, replaced the system whereby the president was elected by an electoral college of 80,000 *notables* with a national election based on universal suffrage, and gave the president of the new Fifth Republic the power to appoint the prime minister and government, to dissolve parliament and to speak for France in international affairs.

Widespread nationalizations immediately after the war and new administrative procedures for implementing the Marshall Plan had given a cadre of modernizing technocrats the levers to guide the French economy, through such mechanisms as Jean Monnet's planning commission. In further

[1] V. Wright, 'France: Recent history and politics', in *Western Europe 1989: A Political and Economic Survey*, Europa, London, 1988.

marginalizing the *notables*, de Gaulle built up the role of these technocrats. He invariably gave the key political posts of prime minister, foreign minister and defence minister to technocrats, thus establishing the single body of technocratic elites which dominates politics and industry and has become a distinctive feature of the French political and economic landscape.

France's postwar economic boom ended in the late 1960s. This period of prosperity, during which France urbanized and became a consumer society, was associated with the 'dirigiste' industrial policies applied with particular vigour by de Gaulle, in pursuit of his goal of restoring France's grandeur. In the early 1980s, the new socialist government enacted widespread nationalizations and re-established state economic planning.[2] This planning function is an important mechanism for an exchange of views between industry and government on current trends and possible future direction for the French economy. The planning commission indicates possible directions for industry but has no direct authority over industrial policy.

Severe economic problems and budgetary austerity led to reduced subsidies for the nationalized firms, forcing them to cut workers and to modernize. An accompanying surge in imports focused the government's attention on the need for French industry to be internationally competitive, as European and world trade became freer. Since the mid-1980s a gradual process of privatization has co-existed with unofficial state direction of private industries. Official influence over many companies survives through large stakes held by publicly owned banks, giving government officials or their appointees seats on company boards throughout French industry.

The French state, therefore, maintains an exceptionally close relationship with industry. In many cases there is no clear separation between the two. It has also accorded industry a very significant political role. This is the background against which modern environmental policy has developed. It is a structure in which the key policy-makers can be unwilling to consider environmental problems seriously, yet one which provides a basis for flexible and innovative approaches to those environmental problems which do receive official attention.

[2] Most of France's largest industrial groups were nationalized, raising the state's share of industrial production from 9% to 24% (ibid.).

6.2. Environmental Policy-making

Legislation on all aspects of the environment is drafted by the appropriate ministry, usually the Environment Ministry, in which case the Directorate for Prevention of Pollution and Risk will be in the lead. This directorate discusses the draft legislation with contacts in other ministries which may be affected, particularly the Industry Ministry. If the ministries disagree over any aspect of the draft legislation after contacts between their relevant directorates, the issue is taken up by special advisers within the 'cabinets' of the ministers concerned. In the event that the special advisers fail to reach an agreement, the issues may be put directly to the ministers and ultimately may have to be resolved by a specialist committee within the prime minister's office.

Officials' perceptions of their working relationships on environmental issues vary from one ministry to another. Industry Ministry officials express the view that their relationship with the Environment Ministry is very cooperative and relatively free of disagreements. The Environment Ministry appears to take a less sanguine view. In private, its officials express frustration with the political strength of the Industry Ministry, giving the impression that it is willing to use its power within the administration to block environmental initiatives.

Once environmental policies have been approved by parliament, various directorates within the Environment Ministry are formally responsible for implementing the legislation. The Industrial Environmental Service (SEI), for example, is ultimately responsible for administering the system of industrial licensing (see below).

The government of prime minister Balladur, which was formed following the defeat of the socialists in 1993, had planned to reform the roles of the Environment Ministry, regional directorates for the Environment and the local *départements*, to give increased decision-making powers to the local and regional authorities. In addition, Balladur proposed administrative measures to improve cooperation between the national ministries on environmental issues, lending weight to the more pessimistic assessment of inter-ministerial relations reported by the Environment Ministry. However, by 1995 there was still no sign of these changes being put into practice.

France does not have an overall plan for the environment. At the initiative of former prime minister Edith Cresson, some environmental issues were included within the five-year planning cycle which the government typically uses to set priorities in major policy areas such as revenue and expenditure, welfare, industry, etc. However, this initiative was effectively forgotten at the moment it was issued. Before that, Bryce Lalonde, an Environment Minister, launched a strategic Green Plan setting environmental priorities and objectives and some measures for central government, but not addressing the role of local government, business or communities.

One area where the Industry and Environment Ministries seem to be in complete agreement – at the official, working level rather than the political level – is the issue of sustainable development, often a major bone of contention in other countries. Environment officials feel that the subject is too vague and too 'top-down' to lead to any specific, useful policy prescriptions in France. In the Industry Ministry, officials confirm this general preference for specific, detailed policies, which give industry a sense of certainty and allow public and private research to be coordinated. They believe this approach to be difficult to reconcile with the concept of sustainable development. If sustainable development has a role it is solely in relation to developing countries and relationships between developing and developed countries.

6.3. The Industrial Licensing Regime

6.3.1. General Description

In France, the system of licensing of industrial sites is administered directly by central government, through regional offices. The pollution inspectorate is based within the Industry Ministry's network of 22 Regional Boards for Industry, Research and the Environment (DRIRE).[3] Officials within the DRIRE offices are employed on the full range of liaison activities with industry. Their duties on administering the environmental licensing scheme are equivalent to 700 staff working permanently on these issues. Sites which

[3] Until recently, the regional network was simply known as 'DRIR', i.e. without the 'environment'. The change of name is a reflection of the government's wish for environment policy to be better coordinated.

require licences are known as *installations classées*, or scheduled installations. A separate organization, DIREN, established in 1993, licenses industrial emissions to water and is wholly answerable to the Environment Ministry.

Typically, proposed regulations are drafted without any *formal*, prior consultation with the industries they will affect. However, the Industry Ministry, which plays a key role in the drafting process, draws on its generally close relationship with industry to reflect that sector's needs. The Environment Ministry receives advice on proposed new regulations from the Conseil Supérieur des Installations Classées. This body, whose director is appointed by the Minister for the Environment, is composed of members of the national inspectorate, and representatives of industry and environmental groups. Regulations, and the standards they contain, do not vary regionally.

The Environment Ministry, in liaison with other parts of the government, uses the BATNEEC principle when establishing emissions standards. However, in many areas, France is waiting for the European Union to establish definitions of Best Available Technology. France was largely responsible for persuading the EU to develop Europe-wide technical BAT guidelines for various industrial sectors. These guidelines are drafted by working groups of national officials, EU Commission officials and industry experts, and approved by the Commission and government experts from the member states. The BAT notes are not in any way binding on the member states, but by incorporating them into its system of industrial licensing, France ensures that it is in compliance with various pieces of EU environmental legislation, such as the Air Directive, which sets Europe-wide standards for emissions of a range of pollutants produced by a number of industries.[4]

However, the EU process has been rather lethargic with the result that – by accident or design – the French regulatory regime is in many areas rather

[4] The European Commission continues to develop its range of technical notes – guidance for the steel, pulp and paper industries was being prepared in 1993–4 – but is increasingly having difficulties committing adequate resources to this task. This problem will intensify when the draft EU Directive on Integrated Pollution Prevention and Control (IPPC) is finally adopted. This will set out a framework for licensing of a greatly expanded range of industries and the Commission is unlikely to take on the task of specifying BAT emission levels for every sector, instead leaving this to member states and concentrating on ensuring that different national implementations are broadly equivalent.

conservative by comparison with its counterparts in other EU countries and with the other countries in this study.

6.3.2. Local Subsidies: A Political Pressure Valve?

Regulations, or decrees, set national emission standards which individual firms cannot legally be compelled to improve upon, regardless of local environmental conditions or strength of feeling. However, a system of investment subsidies has evolved which provides the flexibility for firms to improve upon nationally defined standards, when this is politically desirable.

If a local authority feels that it has a particular environmental problem due to emissions from an industrial site, or if local public opinion is particularly strong regarding those emissions, the local authority may arrange to subsidize the firm's investment in improved or additional equipment. This investment subsidy is funded from local authority taxes and is given, according to the Ministry of Industry, on condition that it does not 'distort competition'. Unlike the subsidy schemes operated by some other countries, there is no requirement for the technology covered by the subsidy to be innovative. However, the fact that the technology is of a higher standard than the prevailing BATNEEC standard is taken as sufficient proof that the subsidy will not distort competition. Inspectors from the local DRIRE office advise the firms and local authorities on those technologies which are appropriate for receiving these subsidies.

This subsidy scheme is a further reason for the national industrial licensing system being less comprehensive than it is in many other European Union countries. Public concern over environmental standards at a particular industrial site often escalates until it is picked up by environmental pressure groups which identify a generic, national environmental threat. The French approach localizes environmental concerns and relieves public pressure for action before they have a chance to become a national issue. Locally, however, pollution can be a very powerful political issue and, according to DRIRE, regional elections have been won and lost on environmental issues alone.

6.3.3. Consequences for Innovation

The French industrial licensing regime is less comprehensive than those in other countries (e.g. Germany, Denmark), encourages local variations as an alternative to raised national standards, and exists within a political structure which gives relatively low priority to environmental issues, at least by comparison with industrial objectives. This is likely to reduce the strategic incentive for firms to invest scarce resources in innovation: polluters can instead influence political processes to their advantage and third party manufacturers face an uphill task in influencing regulations to take account of new price/performance standards. When higher standards exist only locally, this limits the potential market for new technologies or processes, and is likely to favour 'end-of-pipe' responses over investment in innovation.

At the national level, Industry Ministry officials confirm that, in those sectors where France has a strong environmental technology industry (e.g. treatment of urban wastes and waste water), manufacturers concentrate on meeting the tighter standards in their export markets. Conscious of the close political links between polluting industries and the Industry Ministry, and of the export assistance they themselves receive from the government, the technology suppliers have not made any concerted efforts to persuade French regulators to raise national standards.

6.4. Industry, Policy-makers and Regulators: Who's in Charge?

The relationships between industry, policy-makers (responsible for determining the political importance of environmental issues) and regulators (responsible for pursuing agreed environmental goals) are complex, but they provide the key to understanding the scope of French environmental policy and the role of innovation.

The introduction to this chapter describes the conditions which led to the evolution of the technocratic and administrative elites which retain considerable influence over the direction of the French economy and exercise a direct decision-making control over large sectors of French industry. These elites typically have a common educational background, with their careers beginning in the civil service and often holding positions on the boards of major public and private firms. Although the degree of public sector ownership of industry is much diminished from its peak, many of the earlier attitudes and systems of control associated with that time persist today.

As noted earlier, the Industry Ministry has much greater political power than the Environment Ministry even though its influence has declined markedly over the last two decades. The former uses its inside knowledge of industry, with which its natural sympathies are aligned, to play down environmental concerns and pursue a conservative line within the administration. However, the close association with industry which colours the political representation of environmental issues is turned to advantage when the details of regulatory mechanisms need to be thrashed out.

This close association can be traced back at least as far as the immediate postwar period of rapid growth, when the bureaucrats wielded a direct, hands-on control over industries, in many respects running them from the ministries. The cadre of technical bureaucrats, many of whom graduated from the Ecole des Mines, saw this as their natural role and judged themselves to be very good at it. (By contrast, Britain's nationalized industries remained in charge of their own day-to-day commercial affairs, not least because the predominantly arts- and humanities-trained 'generalists' who ran the British machinery of government had little capacity and even less desire to take on technical and managerial functions.) In many ways the French government's procedures for dealing with industry are unchanged from this time, particularly within the Industry Ministry.

For example, when new regulations from the Environment Ministry or the European Union are being developed, the economics service within the Industry Ministry will typically evaluate the likely costs and impacts on French industry. Confident that its personnel are at least as knowledgeable about the effects of environmental measures as companies' own planning departments, the Ministry will not consult directly with industry until the proposed regulations are near completion (although sometimes it will commission independent studies, e.g. from the Ecole des Mines).

Nevertheless, a good flow of information between the ministry and industry is a prerequisite for the ministry to develop the regulations in the first place. Any company which tried to mislead the Industry Ministry over any aspects of its commercial activities or technical achievements, perhaps in the hope of influencing regulations, would provoke a 'big scandal', according to one official. The Industry Ministry would regard this as a breach of trust, running counter to the 'transparent' relationship between industry and government.[5]

Clearly, industry gains from this open relationship. Because the boundary between public and private sector was non-existent or blurred for so long, practitioners have come to see sharing of the details of their commercial affairs with bureaucrats as a natural feature of French business. In return, they expect, and receive, a robust defence of their interests both domestically and, especially, internationally. (There are signs, however, that the special relationship between industry and government – born of the perception of a special role for France – is coming under strain, not least because of the mounting pressure from the European Commission's pursuit of competition policy.)

Industry's close relationship with the Industry Ministry undoubtedly brings a strong political influence which may take the edge off environmental policy. However, this relationship also provides a basis for the ministry to develop flexible regulatory mechanisms which accommodate innovation. Once an environmental goal has been agreed politically, the Environment Ministry relies heavily on the Industry Ministry's expert knowledge in drawing up regulations. The latter will generally be aware of current best practice in the industry, and also of near-term and speculative technologies, costs and time required for these options and the industry's true financial shape. The Environment Department uses this information to draw up its regulations and subsequent BATNEEC guidelines. Because the regulations are applied by the Industry/Environment Ministry officials within DRIRE, changes to the BATNEEC guidelines are applied immediately. In this way, information on the 'state of the art' is disseminated nationally, avoiding the conservative interpretations – and technological lock-in – which can become established in the locally run licensing regimes of Germany and the Netherlands.

With this background of close working relationships between industry and regulator, one might suppose that the voluntary agreement approach being

[5] In other political or commercial contexts 'transparency' usually implies that a relationship is bounded by clear rules of procedure which are in the public domain. The relationship is transparent in the sense that it is open to public scrutiny. However, for the French Industry Ministry, 'transparency' means that officials and their counterparts in industry (often government appointees) have no secrets from one another. The public must simply assume that officials are acting in the national interest.

developed in some other countries would translate well to France. In fact, current thinking is not well disposed towards this kind of regulatory instrument. French licensing authorities have prior experience of industry–government agreements. In the early 1970s, regional water agencies persuaded many of the worst polluters to improve, in the absence of regulations, by offering financial help with pollution control measures.[6] The Environment Ministry, hoping to emulate this strategy, attempted to set up voluntary agreements with various industrial sectors, such as cement manufacturers. Some of these agreements were successful while others were judged to be failures. In the opinion of one Environment Ministry official, the overall outcome was poor and still colours the government's attitude towards voluntary agreements.

Recently, however, the Environment Ministry has returned to the topic once again, and begun to consider whether agreements would be suitable instruments for achieving reductions in CO_2 emissions. In addition, the Commission on Environment, Quality of Life and Growth, which contributed background material to the latest five-year plan for the French economy, suggested that the government should 'favour voluntary agreements, which can mitigate [emissions] more efficiently than regulations or taxes'.[7] Running counter to this view, a recent French theoretical study, primarily of tax measures to reduce CO_2 emissions, was scornful of the potential of voluntary agreements, or 'contracts' in their terminology, to achieve this goal. In particular, the study contended that:[8]

- 'contracts' do not minimize the costs of a given reduction in emissions, whereas a tax approach would;
- the level of emissions associated with producing a good would not be reflected in its price and would not encourage price-sensitive consumers to reduce their consumption, or switch to less CO_2-intensive goods;

[6] Water agencies for each major river basin were responsible for raising water charges. The basin committees, composed of representatives of central and local government and industrialists, could give subsidies to firms taking part in agreed environmental programmes.
[7] Commissariat Général du Plan, *Croissance et Environnement: Les Conditions de la Qualité de la Vie*, La Documentation Française, Paris, 1993, p. 211.
[8] Reported in F. Garcia, 'The Economics of Energy Savings', in N. Steen (ed.), *Sustainable Development and the Energy Industries*, Earthscan, London, 1994.

- the dynamic impact on technical change would be slight;
- initial distribution of targets between industrial sectors would be difficult.

The practical efforts to introduce wide-ranging systems of agreements in the Netherlands and Denmark may soon shed some light on the validity of these conclusions. In the meantime, it appears that French policy-makers and economists are sceptical of agreements which are not accompanied by workable sanctions, i.e. on the extent to which an effective mechanism can be 'voluntary'. Industrial interests are clearly in favour of any system of agreements being strictly voluntary.

6.5. Environmental Technology Strategy

Government support for technology development is an important part of French efforts to meet environmental goals. The Industry Ministry assesses the potential for research and development to reduce compliance costs, supports appropriate R&D programmes and ensures that regulations have the flexibility to accommodate the new technologies as they are developed. Given the general picture of French industrial policy that has emerged so far – of a long-standing obsession with building industrial champions – it is perhaps no surprise that the environmental technology R&D strategy is oriented as much towards exports as to meeting domestic environmental goals. In this respect, the French approach has similarities with the orientation of the Danish R&D effort on environmental technology.

The main organization responsible for funding and coordinating R&D on environmental technologies is Ademe – L'Agence de l'Environnement et de la Maîtrise de l'Energie. Ademe was created in 1990 out of the fusion of three separate agencies responsible for energy conservation, waste and air quality. It is under the joint direction of the Ministers of Research, Environment and Energy. The Minister for Energy is located within the Industry Ministry. This arrangement led to political infighting in the first few years of Ademe's existence, culminating in the resignation of its president and a clarification of the roles of president, now responsible for setting priorities and general policy, and Managing Director, now responsible for day-to-day operations.

Despite its difficulties, Ademe has elucidated a clear conceptual model of the role of technology development in meeting environmental objectives, and the appropriate areas for government support. It has determined that its role in influencing longer-term issues in global pollution and local ecology is limited, as the priorities for these areas are basic investigative research and associated technologies. Ademe would, however, take a legitimate interest in commercial equipment developers in these markets.

Ademe's approach is illustrated by its support for technologies required to accommodate the rapid run-down in CFC production imposed by international agreements. The agency invited CFC users to help it to draw up a research strategy. One group, the manufacturers of large refrigeration equipment and the associated maintenance industry, advised that developing alternatives to CFCs suitable for use in existing equipment or adapting equipment to CFC alternatives would be very costly. They anticipated a large demand for CFCs to maintain the existing stock of refrigeration equipment, long after CFC production has stopped. As a result, while other countries have concentrated on research into alternative substances to CFCs and on modifying basic refrigeration technologies, Ademe has supported the development of cheap and simple CFC recovery equipment. This may allow manufacturers to continue selling old refrigeration technologies. More significantly, it will allow users to continue using their existing equipment, avoiding the high cost of prematurely scrapping refrigeration units.

Ademe believes that this represents a market niche which other countries are neglecting. The strategy it has adopted combines its conceptual approach to determining the proper arena for government support with its other main concern: export potential and French industrial competitiveness.

Since the recession of the early 1990s, the concern with exports has become more explicit and may become a source of tension between the Environment Ministry, which would wish to see Ademe put the needs of the environment first, and the Industry Ministry which has developed a new set of priorities for research on environmental technology.

A new R&D strategy for the chemical industry illustrates the Industry Ministry's new approach. The first step involves the ministry and the manufacturers identifying those products in which French companies have a large share of the world market. This information is then used to prioritize national R&D efforts. For example, basic toxicology research in universities and government-funded institutes can be focused on issues related to these successful products, reducing scientific uncertainties about their environmental risks, and possible legal liabilities.[9] Any problems which arise then set the agenda for further government-supported R&D programmes to develop solutions, e.g. through Ademe – all with the *imprimatur* of 'environmental' research.

6.6. Conclusions

The close interweaving of government and industry interests, and a tradition of technocratic influence over industry have to a large extent dictated the pattern of industry–regulator relationships on environmental issues. Above all, the French approach is a pragmatic one. Environmental pressures arising from the public are relieved with the minimum fuss possible, as with the arrangements for funding enhanced standards at the local level.

When certain environmental standards must be met, often because of European directives, the French administration's intimate knowledge of industry and the existence of a sophisticated, rational technology policy eases the task. Regulators are capable of tailoring implementation in ways which allow innovation to play a part in industry's response. Substantial publicly funded technology programmes ensure that regulators, or their close colleagues, e.g. in Ademe, stay in touch with the state of the art.

There seems little reason to doubt that, politically, industry has a higher priority than the environment, particularly where the national interest (in this case international competition) is concerned. The new strategy of matching environmental research to existing export markets is a potent example. The mood of frustration evident among some at the Environment Ministry supports this view.

From one perspective, France has adopted an ideal strategy. Its citizens' demands for environmental quality seem to be adequately catered for, although they are probably less forceful than those expressed in most other countries in this study. Technology development and innovation are smoothly

---

[9] This involvement of public research organizations has the additional benefit, from the government's point of view, of engaging them in research which is relevant to industry.

engaged when environmental challenges do arise. This strategy will come under strain if either the French public or the international community demands that the government holds the environment in much higher regard. The technocrats would then need to find ways of dissociating themselves from the political interests of industry which they have long championed, while preserving the close communication and understanding which is currently an important means of harnessing innovation. Whatever the political difficulties of such a transition, the supremely competent French technocrats can be expected to handle the mechanics of it with ease.

Chapter 7

Japan

7.1. Introduction

Commentators on Japan's social and industrial successes in the postwar period frequently refer to deeply held values of consensus and cooperation in Japanese society. In the environmental field, these values are generally thought by outsiders to be the basis on which the national bureaucracy, especially MITI, secures the cooperation of industry in meeting environmental objectives. However, although the close MITI–industry relationship is now a distinctive feature of environmental policy, the true strengths of environmental policy in Japan can be traced back to the relationships which were forged between local communities and industry in the 1960s.

The Japanese people have always been conscious of the precarious nature of their existence. Japan is a mountainous country with limited habitable land and is subjected to high rainfall and summer typhoons which cause damage and flooding. Japanese farmers have long been sensitive to the need to farm their land responsibly and cooperate on the sustainable management of the mountains and rivers. In the period of industrialization from 1868 to 1945, it was poisoning of farmland by mining operations which first created public awareness of the harmful effects of industrial growth. For example, at the Ashio Dozan copper mine, about 110 km north of Tokyo, the mine's operators agreed to compensate local farmers for heavy metal poisoning. However, local unrest came to a head in 1900, when violent clashes between police and protesters prompted the government to provide special flood-control measures to prevent further spreading of heavy metals from river sediments onto farmland.

After 1946, land reform transferred property rights to tenant farmers and the new constitution created a system of local government with a high degree of autonomy. Four notorious cases of pollution were to galvanize local communities into finding new powers to protect themselves, in the face of inadequate action from the growth-oriented central government.

In the late 1950s local people began to report the effects of mercury poisoning in Minamata (which gave its name to this form of industrial disease) and Niigata prefecture; of cadmium poisoning in Toyama prefecture; and of sulphur dioxide-related asthma in Yokkaichi. In each case the industrial causes were finally established in the late 1960s and successful compensation lawsuits were concluded in the years 1971–74. But, as a result of these cases, the links between industrial activity and health were clearly established in the public's mind by the early 1960s.

The first response to the new mood of concern was the appearance of environmental impact assessments (EIAs). National and local government offices receiving applications for large-scale industrial projects began to appraise projects' abilities to meet national environmental standards. They issued guidelines on the pollution control equipment necessary to meet standards and outlined requirements for follow-up studies. Another important feature of the EIAs was their requirement for best available technologies, rather than the national standard of best practicable technologies. Grassroots environmental activism also became more evident and effective during this period. Protests by local fishermen and, more significantly, local residents who organized under the slogan 'No More Yokkaichis', caused the cancellation of a petrochemical project in Shizuoka prefecture in 1965. This was the first instance of a mass environmental movement halting an industrial development.

Local governments soon began to impose conditions on new industrial developments which went beyond the existing national clean air and water legislation. Under the constitution, local governments cannot promulgate standards without first being given authority through national legislation. However, they made clear to firms that planning applications might not be approved unless adequate agreements on environmental standards could be concluded. The first such agreement followed an application in 1964 to build a coal power plant in Yokohama. The Socialist mayor was opposed to the project, and set up his own team of experts to conduct an environmental impact assessment. Following this, stringent environmental standards, going far beyond existing national standards, were set out in a contract, which the developer accepted, and the project was allowed to go ahead.

Public activism over environmental matters also had a profound effect on government attitudes. In 1964 the new prime minister, Eisaku Sato, observed 'Because industrial pollution is a distortion in economic growth, it will be strictly corrected'. In the following three years, the Japan Environment Corporation was created, with a remit to help small businesses with the costs of reducing pollution, and the Basic Law for Environmental Pollution Control was enacted. Further air pollution and noise pollution control laws followed in 1968, although many badly affected municipalities considered their provisions to be inadequate. However, a national compensation law made firms liable for any ill-health of citizens in their vicinity, even where there was no clear link to the local firms. The threat of claims for compensation persuaded many firms to cooperate with local authorities in tackling their most obvious emissions, especially atmospheric emissions. Finally, in 1971 the Environment Agency was created, thus establishing the administrative structure which remains to this day.[1]

7.2. Industry and the Local Community

In Japan, major local government bodies – the prefectures and metropolitan authorities – have considerable autonomy. They have played a major role in balancing local communities' fears about environmental degradation with the needs of industry. From this has evolved a flexible regulatory framework which promotes steady technological progress towards better environmental performance.

Japanese prefectures are responsible for ensuring that industries comply with national regulations such as the 'k-value' rules on emissons of SO_2 from boilers.[2] In some areas these national regulations are insufficient to achieve air quality standards. A group of prefectures covering the most heavily industrialized areas of the country have, therefore, been given authority to apply their own standards through local ordinances. This is also a recognition that local action

[1] For a comprehensive review, see Michio Hashimoto, 'The Japanese Experience of Tackling Pollution', *Japan Review of International Affairs*, 7, 1, Winter 1993.

[2] National legislation sets emission limits according to a formula which includes factors such as boiler output, stack height and local climate. An adjustable factor, the 'k-value', introduces flexibility which allows emission limits from the same type of equipment to differ between areas with good and poor air quality.

on environmental issues developed ahead of national legislation in these areas. Local authorities have augmented this power to make ordinances with persuasion and formal contracts. As a result, a subtle trial-and-error regulatory process has evolved, based on close relationships and frequent communication between officials and industry representatives, exchange of detailed technical information and an underlying respect for public opinion.

7.2.1. The Three Tiers of Local Regulation

Within the Tokyo Metropolitan Authority (TMA) area, 10% of the country's population consumes more than 10% of the nation's energy. The TMA, a hybrid of regional prefecture and city authority, occupies an imposing double-tower skyscraper in Shinjuku, giving an impression of power which is probably entirely appropriate. The TMA's strategy for controlling air pollution demonstrates how such power can be acquired and dispensed responsibly. There are three tiers of environmental controls in the TMA strategy. The first is a 'recommendation', the second is a local ordinance and the third is a contract specific to an individual site.

A recommendation is variously referred to as a guideline, as 'administrative guidance' or *'Gyosei Shido'* in Japanese, and is also employed by national agencies. For example, the TMA has issued a recommendation that industrial sources of NOx should produce only half the level of emissions permitted under the national law. This applies to around 7,000 facilities such as incinerators or gas or oil-fired boilers producing power and heat in factories. Existing oil-fired boilers are expected to meet a limit of 110 ppm NOx while new boilers should not exceed 60 ppm. Small variations in the limits are intended to account for burner capacity, fuel type and location. Smaller boilers (less than 10 m^2 heat exchange surface) are exempt, but the recommendation is publicized as a target, along with information on boilers which can achieve this level of emissions. Not all of the larger existing facilities have complied with the recommendation, which was enacted in 1991. However, they have strong incentives to do so, if they wish to avoid being caught out by a sudden elevation to the second tier of environmental regulation.

This second tier is the local ordinance, which has the force of law and is promulgated and enforced by the local authority. After several years of experience with the recommendation on NOx emissions, the TMA will have acquired a greater understanding of the costs borne by facilities and of the level, and pace of advancement, of the technologies. Local authority officials will have made the mistakes which are an inevitable part of any new scheme, in a non-legislative framework and these will have been accepted by both sides as a necessary part of the learning process, and resolved amicably in most cases. Fine tuning of the standards may have occurred along the way and a final opportunity for this is available when the ordinance is drafted. The legislation itself is unlikely to prove controversial during its passage through the TMA legislature since most of the uncertainties will have been resolved in the preceding years. In the Tokyo area, this smooth legislative process has recently been successfully applied to the asbestos removal recommendation which was first introduced in 1991, and was upgraded to an ordinance in mid-1994.

The third tier of local authority environmental standard-setting is the contractual agreement. This is a form of voluntary agreement which local authorities make with companies or groups of companies involved in large-scale developments, which agree to meet environmental standards tailored specifically to each development. These contractual agreements evolved in the 1960s, when local authorities were under intense public pressure to reduce industrial pollution but national laws were unequal to the task. Bodies such as the TMA used their authority over planning applications to persuade firms to accept emissions restrictions which were not required by law, either because national standards were lower or because they did not exist at all. Local authorities routinely exchange information with one another on the nature and success of these agreements and over the years this has created a powerful ratcheting effect on emissions levels across a wide range of processes and pollutants.

7.2.2. Ratcheting Emission Standards – An Example

To understand how this works, we can return to our example of NOx emissions from boilers. The current national standard sets a limit of 60 ppmv for the stack emissions from the largest natural gas fired boilers, e.g. those used in power stations. However, the Futtsu gas-fired electric power station, which stands on the eastern side of Tokyo Bay, meets a much stricter standard. The voluntary agreement which the Tokyo Electric Power Company

(TEPCO) concluded with Chiba prefecture, prior to commissioning of the power station in 1985–88, sets a limit of only 10 ppmv. Continuous emission monitors installed in each of the station's stacks typically show emissions of only 8–9 ppmv, with anything over 9 ppmv prompting corrective action by the station operators.

Obviously, TEPCO was not asked magically to produce technologies which could achieve this six-fold reduction in emissions overnight. This level of technical performance evolved gradually. Previous voluntary agreements at other, similar plants had resulted in the agreed standards being exceeded because of the plant designers' need to build in a sufficient margin of error to cope with poorer than expected performance from any part of the process. More often than not the plants achieved emissions below the agreed level, sometimes up to 40% lower. Other authorities were aware of the new performance levels (not least because the utilities freely offered the information), which became the benchmark during negotiations on agreements for subsequent, similar projects. Over time, this continuous downward pressure on emissions has created large disparities between the national standard and the standard being achieved in practice, as at Futtsu.

7.2.3. Consultation and Conflict Resolution

This three-tiered regulatory structure – recommendation, local ordinance and contractual agreement – is supported by an institutional structure and an information gathering process, both of which are echoed at the national level. In the capital, this institutional structure is centred on the Environmental Council of Tokyo, an advisory body consisting of academics, TMA officials and officials of national ministries such as MITI and the Environment Agency. The council advises on the general measures required to meet Tokyo's environmental quality objectives. If a specific TMA recommendation, such as the NOx standard, seems to be causing technical difficulties, a sub-committee of the Environmental Council consisting mainly of technical experts will be created to examine possible solutions. These might include calling for MITI to target national research funds on the problem, or suggesting that the recommendations should be amended in some way.

TMA officials ensure that they are well informed on the technical issues relating to the sectors they intend to regulate. An important element of this is a formal system of *technical hearings*. In the case of NOx, for example, boiler manufacturers, pollution control equipment manufacturers, fuel suppliers and plant operators all testify on the current state and direction of the technologies. These hearings are held in private and commercial confidentiality is of course respected. This partly explains why firms cooperate fully with the regulatory authorities, almost without exception. More generally, officials will maintain networks of contacts with academia, national government agencies, research bodies, other local authorities, industry associations, etc., in order to stay abreast of their fields.

The intense public activism over pollution issues in the 1960s has prompted the evolution of a locally based system for resolving pollution disputes, institutionalizing the active role taken by many local authorities. The national Law Concerning the Resolution of Environmental Pollution Disputes, of 1970, set out guidelines for prefectures to establish environmental pollution investigative councils. These bodies seek to avoid having disputes end up in court by providing a mechanism for 'conciliation, mediation and arbitration'. In Tokyo, public complaints have subsided from a peak of over 16,000 in 1972 to about 8,000 in 1991, with just under half of these being complaints of commercial/industrial noise pollution and the remainder relating to air, water and soil pollution, odour, vibration and other causes. The investigative council has dealt formally with over 100 environmental disputes in the same period.

Even the intervention, as a result of public concerns, of the municipal authorities in the debate on auto emissions in the early 1970s (see Chapter 9) has been partly institutionalized. The TMA continues to conduct independent surveys and research, particularly on NOx emissions from vehicles. When appropriate, it submits formal requests for tighter regulations to the national Environment Agency and the vehicle manufacturers.

7.3. National Structures

We have seen that national environmental policy in Japan has often been shaped by earlier initiatives at local level, especially where traditional manufacturing and heavy industry are concerned. National standards often lag behind local standards or local recommendations which become *de facto* standards. In all areas of environmental policy, the national parliament, the Diet, plays only a minor role. This tends to be the case in many other policy areas, as the parliamentary system is primarily concerned with propagating itself, almost to the exclusion of all other issues. As a result policy development is largely the preserve of the bureaucracy. The national environmental policies are reviewed below.

7.3.1. Legislation

The Basic Law for Environmental Pollution Control of 1967 established environmental standards for air, water and soil quality and for noise levels, and formed the basis for policies to tackle these forms of pollution, as well as vibration, odours and ground subsidence. This gave rise to the air, water and noise pollution control laws and the dispute settlement and health-damage compensation laws, all of which were passed between 1968 and 1973. Entirely separate pieces of legislation such as the Nature Conservation Law were also enacted during this period. However, the centrepiece of national environmental policy remained the Basic Law. In addition to providing a legal base for environment policy, this set the tone for the relationship between government agencies, local authorities and industry. Major shifts in these relationships, particularly in the late 1980s and early 1990s, are reflected in a recent major overhaul of the Basic Law – the new 'Basic Environment Law' of November 1993.

7.3.2. MITI-gating the Pain

Under the old Basic Law a clear distinction was drawn between formulating legislation and implementing it. However, whereas in the US an elected legislative body – the Congress – makes the law, and a bureaucracy – the Environmental Protection Agency – implements it, in Japan the bureaucracy does both. The Environment Agency leads and coordinates drafting of new laws, negotiating with relevant ministries.

MITI has played a major role in implementing much of the environmental legislation affecting industry. Until recently, it often issued guidelines to industry, in accordance with a law dating back to the 1950s. MITI originally used the Law for the Promotion of Mechanical Industries *(Kikai Shinko Rinji Soti Ho*) to spur key industries to improve to world standards by setting performance standards. The law expired every five years and was replaced with a slightly revised law reflecting new industrial priorities, e.g. electronics in the 1960s and information technologies in the 1970s. These priority areas drew on the latest MITI 'Vision', a long-term forecast of industrial priorities.

MITI was able to issue guidelines on environmental performance under the laws on industrial promotion, often expressing these as types of technology which firms in a particular industrial sector should adopt or develop within a certain time. Firms which ignored MITI's recommendations without good reason faced the threat of their non-cooperation being made public, a threat so dire that it never had to be used.

Inevitably, MITI faced a conflict of interest which led to suspicions that it was not as zealous in setting guidelines as were the local authorities. However, the quality of information MITI elicits from firms, and on which it bases its recommendations, cannot be underestimated.

This has led to occasional conflicts with other agencies. It thus happens that when MITI is consulted on proposed local authority ordinances it may well disagree with the local authority's view, for example on the basis of the short-term R&D and management costs for firms. The local authority will generally need to be able to point to a reasonably successful period of applying the proposed ordinance through guidelines, in order to counter MITI's objections.

MITI has also been suspected of acting as a brake during the early stages of policy formulation. When new national legislation is being developed, MITI, along with other ministries, such as construction and transport, canvasses and represents the views of industry to the Environment Agency. However, some officials report that institutional rivalry has in the past limited the flow of information between the Environment Agency and MITI in particular, and that this contributed to a mismatch between the aspirations of environmentalists when the Environment Agency proposed or announced new legislation, and the reality once MITI began to implement it.

In MITI's case the network of associations which exists between all branches of the bureaucracy, the business community, academia, etc., has been enhanced by a steady flow of personnel from MITI into major firms. Postwar legislation requires that the top stream of civil servants serves no more than 25 years in public office. Until the recession of the early 1990s, MITI officials commonly found their second careers in industry. The firms gained from the close personal relationships these 'retirees' maintained with existing MITI officials and MITI had greater confidence in the information it was given by industry. However, in recent years, many major firms – often facing difficulty finding enough posts for their own senior staff – have been less willing to take on former bureaucrats, forcing many MITI officials to seek second careers elsewhere.

7.3.4. The Environment Agency – the Ministry that Never Was

The Environment Agency was created in July 1971. It was given a horizontal role coordinating environmental policies across the major ministries and proposing (i.e. creating) environmental laws. Within the bureaucracy there is a clear pecking order, in which MITI and the Finance Ministry are the most popular choices for each year's new intake of top-stream officials. These new recruits do not favour the Environment Agency as their home department, largely because it lacks ministerial status. As a result most of the key staff of the agency are borrowed from other ministries for short periods, including many from MITI. The agency's attractiveness has improved a little recently with the rise of global environmental issues such as ozone depletion and global warming. During 1993 there was an attempt to give the agency full ministerial status, but this effort got lost in the general turmoil of political reform following the defeat of the Liberal Democratic Party.

Despite this the Environment Agency has developed a highly successful modus vivendi with industry. This relationship must be seen as part of the bigger picture in which there is constant exchange of views and information between all relevant public and private institutions and representative bodies. The most notable expression of this information flow is – as was the case with the local authorities – the technical hearing system. Technical hearings are employed in those areas where the Environment Agency is responsible for implementing legislation, in addition to its usual role of coordinating its formulation. The origin of the technical hearing system as it applies to vehicle emissions is described in Chapter 9. The success of the system rests on the willingness of firms to share their technical information with the agency. In general, firms are willing to do this because they have learned that the agency will adopt a flexible approach to target setting. For example, it will ask manufacturers whether they can meet a lower NOx standard for trucks. A regulation will be promulgated with suggestive, but not yet binding, future dates for the standard coming into effect. Subsequent regular hearings will determine the exact timing. An emission target is, therefore, regarded as a technical problem to be solved by the firms and backed up by the force of the agency's regulations at the most appropriate time. Standards for vehicle pollution in particular are revised so frequently that this has become a continuous dialogue, with hearings related to the implementation of an existing standard feeding into the process of drafting a new standard.

In summary, environmental rule-making at the national level displays an interlocking set of processes in operation: the Environment Agency's technical hearing system, which ensures that all relevant information is considered prior to legislation and then formally repeats the process, to determine the best time scale for implementation; a similar but less formal dialogue between MITI (and sometimes others) and industry, which allows MITI to determine optimal time scales for implementation, where it has that responsibility; finally, the local authorities, again maintaining a detailed, open dialogue with industry and providing a direct route for the public's concerns to be fed back to firms as an obligation continuously to reduce their impacts on the environment.

7.4. Future Developments: the New Basic Law

The new Basic Law differs from the previous version first in placing sustainable development at the heart of environmental policy. Articles 3, 4 and 5 are titled 'Enjoyment and future success of environmental blessings', 'Creation of a society ensuring sustainable development with reduced environmental load' and 'Active promotion of global environmental conservation through international cooperation'. It creates a legal base for a wider range of policy measures such as education and financial instruments.

A national requirement for environmental impact assessments appears for the first time in the new Basic Law. Previously, local rules such as the Tokyo metropolitan environmental impact assessment ordinance were all that existed in this area. However, many companies are disappointed that local authorities will continue to lay down their own rules for environmental impact assessments, rather than adopting a common, national approach.

7.4.1. Voluntary Agreements

One feature of the new Basic Law which is set to make the relationship between industry and government more transparent is the framework for voluntary agreements on pollution reduction.

Article 8, 'Responsibility of corporations', addresses the pre-existing responsibility of firms to comply with mandated standards on pollution reduction and waste disposal. However, paragraph 4 states: 'corporations are *responsible for making voluntary efforts* to conserve the environment such as reduction of the environmental loads in the course of their business activities'.

Currently, MITI has the lead responsibility for the programme of voluntary measures by industry. This began in October 1992, when the new Basic Law was still being drafted, with a call for industry to present proposals for voluntary pollution reduction. By October 1994, 362 firms including 60% of the major manufacturers with over 300 employees had submitted their plans. These typically include measures such as environmental audits, involvement of local communities, simplification of distribution and waste reduction and recycling. MITI is pressing more firms to draw up plans and is helping firms to improve their existing plans by identifying and promoting the best features of the plans received to date. This helped to raise the number of plans which include numerical targets for pollution reduction from an initial very low base to between 20% and 30% of plans submitted.

In the longer term, it is likely that the programme on voluntary measures will evolve into a comprehensive framework for tackling industrial pollution. MITI has instructed one of its advisory councils to examine possible strategies for 15 major industrial sectors. This will bring a coherence to the overall voluntary measures approach and will draw in many of those companies which have not yet produced their plans.

The voluntary agreement scheme builds on existing relationships and makes them more open and transparent. The contractual agreements which have long been concluded between local governments and industry can now follow a nationally agreed system which might act as a benchmark for the efforts which firms should make. At the national level, the voluntary agreements give firms the initiative in proposing environmental measures and should therefore be seen as a strengthening of the processes of communication which MITI draws upon when setting guidelines or recommendations.

7.4.2. A Basic Plan for the Environment

A requirement for a Basic Environment Plan is laid down in Article 15 of the new Basic Law. The plan will set out the 'comprehensive and long-term policies for environmental conservation' and 'the matters required to comprehensively and systematically promote' these policies. The Central Environment Council of the Environment Agency, an advisory body composed of around 40 representatives of academia, industry and the media, is responsible for drafting the plan. This task was expected to be completed by the end of 1994. The plan will oblige local authorities in areas with serious pollution problems to draw up an environmental pollution control programme, i.e. a local version of the Basic Plan. In practice, these areas will be the heavily polluted ones, such as Tokyo, which currently have autonomy to make their own pollution ordinances. The local programme will be approved by the prime minister, on the basis of advice from a new body, the Conference on Environmental Pollution Control.

Under this new structure, local authorities are likely to retain their existing powers to set local environmental standards. However, the Basic Plan will set limits on regulatory actions to combat CO_2 emissions, and perhaps on other issues of national and global importance. A local authority would risk having its programme rejected by the Conference on Environmental Pollution Control, if it appeared to go beyond the Basic Plan. It is unlikely that this possibility will seriously erode the credibility of local authority regulators in the eyes of polluting firms, as long as the local authorities maintain the freedom to act on 'traditional' pollution and waste issues.

Assuming there is little effect on the autonomy of local authorities, the Basic Plan will have its greatest impact in creating a national climate which will encourage a shift towards voluntary measures by industry. Its very existence will promote environmental protection as a nationwide goal, which citizens and industry alike are expected to contribute to. This will make many more firms aware of their obligations to take voluntary actions to adopt best technologies and develop cleaner processes. The emphasis on a long-term, comprehensive approach, to be backed up by any necessary measures, should convince most firms that whatever their current legal position, they would be better off dictating their own pace of environmental improvements. This nationwide recognition and adoption of voluntarism, now backed up by a coherent national strategy, build on the successful model of industry–regulator cooperation which has evolved over decades at the local level.

7.4.3. A Carbon Tax?

The final major innovation of the new Basic Law is its provision for a wider range of policy instruments to be used to achieve environmental goals. Article 9 makes citizens responsible for making efforts 'to reduce the environmental loads associated with their daily lives' and 'to conserve the environment'.

Article 22 introduces economic instruments, and was included almost exclusively to provide a legal base for a carbon tax, at a time when it seemed likely that action in Europe or the US might oblige Japan to follow suit. During drafting of the Basic Law, MITI was, and remains, sceptical about the merits of a carbon tax. MITI argues that since Japanese industry is more energy-efficient than most of its counterparts there is little scope for significant energy savings as a result of higher energy prices, at least in the short term. The tax would simply drive up costs, encourage some firms to relocate in less energy-efficient economies in south-east Asia, where there is scant prospect of carbon taxes, and so produce a net increase in CO_2 emissions. (This was also one of the main arguments deployed by the United Kingdom in justification of its decision not to support a European Union carbon/energy tax.)

Of course, industry ministries world-wide have objected to the possibility of carbon taxes, but MITI's concern is particularly well founded. It has long experience of directly influencing energy use in industry, through the Energy Conservation Law, and therefore a solid basis for arguing that a tax might be relatively inefficient in Japan. Officials suggest that a carbon tax would do little to increase the pace of development of more energy-efficient technologies and processes. They fear that the tax will be an unavoidable drain on firms' financial resources which will reduce their capacity to adopt or develop technology in response to the detailed sector by sector and plant by plant guidelines favoured by MITI.

Japan's Energy Conservation Law

Japan introduced the Energy Conservation Law in 1974, in response to the first oil crisis. It was last amended in 1993. Under the law, an industrial site with annual fuel consumption of 3,000 tonnes of oil equivalent or more, or annual electricity consumption of 12 GWh or more, is a 'designated' site and must appoint a qualified energy manager. In 1989 there were over 4,600 such sites, each reporting directly to a MITI official, and over 40,000 qualified energy managers. Guidance to firms is detailed and specific. MITI sets guidelines for the energy required to produce a unit of output, based on best practice within each industry and provides more detailed information, e.g. on the ideal air/fuel ratio in a particular type of boiler.

MITI can insist that a factory which is lagging behind the industry average updates its equipment or processes. Officials will take account of factors such as current profitability and the time since the last relevant investment, and may arrange for preferential loans or tax relief, as provided for by the Energy Conservation Law. If a firm does not comply, MITI can make a public announcement that its performance is inadequate and make a legal order forcing the firm to take remedial measures. Firms which ignore an order can be fined. In practice the threat of a public announcement is invariably enough to ensure compliance. Smaller, non-designated companies are provided with information, advice and financial assistance to reduce their energy use, and awards are made annually to suppliers of novel equipment with reduced energy consumption.

Other measures are aimed at consumers and manufacturers. Energy labelling has been compulsory for air conditioners, cars and fridges since 1979. More recently, televisions, fluorescent lamps and photocopiers were added to the list. (MITI also sets efficiency standards for these products which manufacturers must adhere to.) An official energy conservation day and energy conservation month help to drive the message home.

7.5. Industrial Organization

The distinctive organizational structure of the Japanese business community has made an important contribution to the relatively smooth progress of environmental regulations and allied technological progress. To many non-Japanese, the most distinctive feature of the Japanese business world is perhaps the *Keiretsu* – large groups of firms which are linked by long-term supply contracts, cross-ownership of shares, or exchange of personnel. A *Keiretsu* typically has only one firm in any given industry, and competes fiercely with other *Keiretsu*. They exist in three main forms – distribution, bank-centred and, of most interest here, production Keiretsu.

Production *Keiretsu* are headed by one major manufacturer, capable of providing technical, managerial and financial assistance to a host of major contractors and sub-contractors. They are conducive to innovation in several ways. They allow for risky investments, e.g. R&D or innovative technologies, to be shared and provide a testing ground for new products and practices within sympathetic customer–contractor relationships. Large firms within a *Keiretsu*, for example the first and second tier component suppliers in the vehicle industry, can be a source of information and advice for their smaller sub-contractors on how to meet environmental standards, so promoting uniformity of achievement of standards within Japanese industry. The larger firms also act as a channel of communication in the other direction, relaying the views of their smaller *Keiretsu* partners, which might only have a handful of employees and little time for networking, to local and national regulators. This helps to avoid the severe difficulties regulators in many other countries face in determining the needs, capabilities and performance of small and medium-sized firms.

Another distinctive feature of Japanese business, especially in large companies, is the commitment to lifetime employment. Despite the pressure of the recession of the early 1990s, most Japanese manufacturing companies have resisted complete abandonment of this principle. Some Japanese business leaders believe that lifetime employment is helpful to innovation, as it allows cross-fertilization between departments as staff move from one to another, allows staff to concentrate on and receive recognition for long-term improvements, and fosters trust and hence communication between top management and those engaged in production, where most innovation arises. These benefits also help Japanese firms to understand and respond to environmental problems, which are by their very nature long-term and cross-functional and require innovative solutions.

7.6. Conclusions

Japanese environmental policy has been highly successful in cutting down on the severe industrial pollution which blighted cities in the postwar period of industrialization. Yet the very tough regulatory regime which has accomplished this has not produced the protestations of excessive cost and regulatory burdens which accompany German or American regulatory efforts.

Some of this can be put down to cultural differences. The postwar industrial culture, where the firm is seen as a community and as a part of the local community, has given most Japanese firms a strong social conscience. A general respect for authority is evident throughout Japanese society. But cultural differences are not the only reason for the low level of friction over environmental policy in Japanese industry. There is a genuine difference in perception of environmental issues, stemming from both the context and the conduct of environmental policy.

Local authorities have taken responsibility for much environmental policy and have helped to dictate its national political context. This is not automatically a good thing, but in Japan it has had the effect of decoupling environmental politics from national strategic industrial policy. Environmental policy, towards industry especially, has therefore been consistent and politically stable with the emphasis clearly on protecting the health of local citizens. As in Germany, industry has learned that it will not get away with polluting activities which come to be of concern to policy-makers.

Just as important, and complementary to the stable political context, the mechanisms through which policy is conducted are based on open, informed dialogue between policy-makers and firms. Widespread use is made of 'technical hearings', where polluting firms and technology suppliers share their technical and commercial knowledge with regulators and find a consensus on the prospects for innovation. These technical hearings have been described as simply an outward manifestation of a multilateral flow of information between local and national agencies, individual firms, citizen groups, chambers of commerce (which are places to exchange practical experience rather than political lobby groups), academia and quasi-public research bodies. This is where Japanese experience departs from the German experience, where the dialogue is poorer. Japanese firms know that there is no escaping environmental responsibilities but they also know that the solutions will take their needs and capacities into account and should not be too costly and painful.

Competence of the policy-makers to understand industry is an essential part of making this dialogue work. However, firms must also be competent in dealing with innovation internally, to enable them to communicate their abilities and limitations to policy-makers. In Japan industry's internal capacity for innovation and ability to manage this may be better than in many other countries. The business structure typified by the *Keiretsu* helps to support innovation in smaller firms. Small firms gain assistance with organizational complexities of managing innovation; profit sharing agreements can allow small firms to benefit from innovation more than they would in an adversarial supplier relationship; and the large firms can help smaller firms to adopt new technologies and techniques which have been developed elsewhere. A widespread culture of continuous improvement, or *kaizen*, leads directly to reduced pollution (see Chapter 2).

For the regional and global environmental challenges of the future, the Environment Agency is developing national policies which draw on the local authorities' approach to policy-making, i.e. industry-led voluntary action within a coherent long-term strategy.

Chapter 8

The United States

8.1. Introduction

The concern of Americans for their environment first found political expression around the turn of the century and led to conservation-related legislation on national parks, fish and game management and migratory bird protection. Later, growing recognition of the harmful effects of economic activity led to federal legislation which specified local responsibilities for controlling pollution sources, such as the Water Pollution Control Act (1948) and the Clean Air Act (1963).

Modern environmental policy began to take shape in the late 1960s and is often said to have been born on the first Earth Day, in 1970. Environmental activism was closely associated with movements opposed to the Vietnam war and was a means for the new generation to express its disillusionment with the values of the preceding one. It is not surprising that environmental politics quickly became adversarial. The inherent polarity of the environmental issue, which causes difficulties for most societies, was magnified in the United States by this difficult birth and by the generally adversarial working practices of American political, administrative and judicial institutions.

It is no coincidence that, of the countries in this study, arguments and counter-arguments about the excessive cost and burden of environmental regulation are strongest in the United States. This chapter illustrates how the adversarial, legalistic approach to environmental issues has produced an inflexible, fragmented and confused regulatory system, which stifles innovation and so frustrates industry that opposition to environmental goals seems preferable to seeking creative solutions.

Policy-makers and industrialists have recently begun to realize these problems. Experiments with new procedures aimed directly at encouraging innovation are under way, especially in California where industry's concerns over regulation reached a high point in the late 1980s. More fundamental reforms, capable of altering the political context of environmental issues, will need to focus on national institutions.

8.2. The Development of Modern US Environmental Policy

Political and public awareness that overall environmental quality was a national and even global concern had grown steadily throughout the 1960s. Rachel Carson's book about the impacts of pesticides on wildlife, *Silent Spring* (1962), and *The Quiet Crisis* (1963), written by the Secretary of the Interior, Stewart Udall, were both best-sellers. Presidents Kennedy and then Johnson sent special messages on nature conservation to Congress. This 'new conservation' was displacing the old definition which had given as much emphasis to conserving the capacity for economic growth through sustained exploitation of natural resources, as it had to setting aside natural areas and protecting species.

When the Nixon presidency began in 1969, this change in emphasis was so well established that Nixon's nominee for Interior Secretary, Walter Hickel, a former Governor of Alaska who equated natural resources with opportunity for economic development, was nearly rejected by the Senate Interior Committee. Only the appointment of a highly regarded environmentalist as Under Secretary of the Interior saved Hickel's nomination.

Nixon signed into force the National Environmental Policy Act (NEPA) in 1970 and presided over the 1970 Clean Air Act (CAA), which set ambitious national goals for air quality. At the same time, a major reorganization took place within the executive branch of government (the collective name for the President, his White House staff and advisers, and the departments and agencies corresponding to ministries in other countries). The CAA departed from previous environmental legislation by requiring federal agencies to base their decision-making on scientific analysis of ecological impacts. It required the administration to determine national air quality standards and develop regulations to enforce those standards. This new requirement for interdisciplinary scientific analysis, as a precursor to federal activities and rule-making, prompted the creation of a dedicated federal agency, the Environmental Protection Agency (EPA).

Congress was concerned about the possibility of 'agency capture' of the EPA, i.e. the new agency falling under the influence of the parties (mostly industry) it was trying to regulate.[1] Legislators guarded against this by framing the enabling legislation in a way which biased the administrative structure and staffing towards environmentalists. Another tactic Congress used was the 'hammer clause'. These were clauses within the primary legislation, such as the CAA and the Water Pollution Control Act (1972), which gave the EPA very strict compliance deadlines. 'Soft hammer' clauses were originally framed to ensure that regulations should be developed by a certain date. Later, 'hard hammer' clauses were employed. These spelt out measures which would automatically become law if the EPA had not developed its own regulations by a certain date. Congress hoped that this would reduce the ability of those agencies which were pro-industry, such as the Commerce Department, to delay or block the EPA's regulatory, rule-making activities.

Another tactic used by Congress to pressurize the agencies was the so-called 'technology-forcing' requirement. The 1970 CAA forced the EPA to develop technology standards which would achieve certain levels of emissions from vehicles by 1975. These provisions invariably included get-out clauses (on the grounds that Congress could not say with certainty how technologies would develop) which were exploited by the special interests the legislators were trying to resist. One of the most effective measures used by Congress to thwart agency capture was to provide citizens and citizens' organizations with legal 'standing', i.e. the right to sue federal agencies which were not meeting their statutory obligations under the various environmental acts.[2]

The approach of Congress in the 1970s institutionalized confrontation between regulator and regulated. This reached its ultimate expression during the Carter administration, when the legislators responsible for the Surface Mining Control and Reclamation Act (1977) included the provision that

---

[1] There is a general tendency in the US to exclude business people from regulated areas they have previously worked in, to avoid possible conflicts of interest.

[2] See, for example, Valerie M. Fogleman, 'Economic Impacts of Environmental Law: the US Experience and its International Relevance', in Nicola Steen (ed.), *Sustainable Development in the Energy Industries: Implementation and Impacts of Environmental Legislation*, RIIA, London, 1994.

prevented the new implementing agency, the Office of Surface Mining Reclamation and Enforcement, from hiring anybody associated with 'any legal authority, program or function in any Federal agency which has as its purpose the development or use of coal or other mineral resources'.[3] In effect, no public servant with any knowledge of the mining industries was allowed to serve in the new agency.

Industries were soon crying out against the costs of over-ambitious environmental goals, implemented in haste due to hammer clauses and in an insensitive and inefficient way by regulators with no experience of industry. In 1980 the newly elected President Reagan responded to the calls for 'regulatory relief' and the pendulum swung rapidly to the other extreme. Massive cuts in the EPA's budget reduced its staff by a quarter. Anti-environment ideologues were appointed to key positions. Interior Secretary James Watt suggested that the world need not worry about natural resources running out because the 'biblical Armageddon' would occur first.[4] Anne Burford, the EPA administrator, exploited any vagueness in the primary legislation to block regulations and faced numerous hostile, public interrogations by Congressional committees. Both were dropped in favour of less controversial figures prior to the 1984 presidential election, as Reagan tried to lower his increasingly unpopular anti-environment profile.

One of Reagan's most effective blocking tactics was the use of cost–benefit analyses for all proposed regulations. This ensured that draft regulations disappeared into a 'black hole' in the Office of Management and Budgets, even when the primary legislation, for example the 1970 CAA, explicitly stated that costs could not be included as a criterion in formulating regulations. Environmental policy in the 1980s settled into a pattern where Congress, dominated by Democrats, drafted increasingly detailed legislation in order to constrain the EPA, and the president habitually vetoed these acts.

The Clean Air Act is the prime example of this process. Congress began work on updating the CAA in the early 1980s but it was not until 1990, during the Bush presidency, that it was finally adopted. By this time it had

---

[3] US Public Law 95–87 [1977], Sec. 201 (b).

[4] Quoted in Henry P. Caulfield, 'The Conservation and Environmental Movements: An Historical Analysis', in James P. Lester (ed.), *Environmental Politics and Policy*, Duke University Press, Durham, NC, 1989.

grown out of all proportion. One extraordinary passage illustrates the obsessive lengths that the Congressional committees went to in order to force the EPA to do their bidding. This sets out the limits for emissions of sulphur dioxide (SO_2) from coke ovens, and specifies individual maximum percentages of emissions from the doors, door seals and top seals. It is extremely unlikely that these emission ratios would correspond to state-of-the-art technology nearly ten years after they were first drafted, regardless of the quality of the technical advice on which they were originally based. More importantly, it is a cause for concern when the legislators of the world's greatest military and economic power feel compelled to spend their time on such detailed rule-making.

The Bush presidency undoubtedly marked a turning point in American environmental politics. The new EPA administrator, William Reilly, seemed genuinely committed to the environment. On the other hand, environmental regulations were a major target of Vice-President Dan Quayle's Council on Competitiveness which took the view that they were invariably a burden on industry. Despite this legacy of Reaganism something had changed fundamentally. Once again the 1990 Clean Air Act Amendments provide the best example. Certainly the act contains the ghosts of confrontation but it was accepted by Bush because of its policy innovations, particularly on measures for tackling acid rain. Emissions of SO_2 from power stations are to be controlled through a market in tradable emission permits, in line with advice from economists that market instruments should generally be less costly than technical and emission standards. Electric utilities are compelled to install continuous emission monitors and ensure that they have sufficient permits to cover their emissions, under threat of fines. However, they are free to choose how to reduce their emissions, without interference from federal, state or local regulators.

When the new Democratic President, Bill Clinton, took office in 1992 it soon became clear that his administration intended to build on this new approach. After an initial period of institutionalized confrontation between environmentalists and industry in the 1970s, followed by a reactionary, laissez-faire anti-environment stance in the 1980s, it seems that American environmental policy is attempting to evolve the pragmatic, cooperative relationships between industries and regulators which are apparent in other countries. The remainder of this chapter fleshes out the details of this evolving relationship, at federal and state level, and discusses the substantial conservative forces which may frustrate a successful outcome.

8.3. Major Institutions in US Environmental Policy

8.3.1. Congress and the Executive Branch

The American constitution was designed to maintain a balance of powers between the executive branch, led by the president; the legislative branch, composed of the Senate and the House of Representatives which together make up the Congress; and the judicial branch. Early in the country's history, new legislation was approved only if all bodies agreed that it was in the national benefit. Since the Second World War, securing funding for election campaigns has become increasingly important for Congressmen seeking election or re-election. Where they were once relatively free to vote according to their consciences (or to respond to whatever party discipline there might have been), now they must think of the war-chest required to purchase air time on television. With many candidates dependent on fund-raising efforts, and thus on their major campaign contributors, 'special interest' politics has flourished.

Special interest politics has positive and negative expressions. It is positively expressed when Congressmen attach pet projects to any convenient though unrelated piece of legislation that happens along. The negative expression arises when they oppose any legislation which might harm a campaign contributor, or other special interest with influence, even if the measure is clearly in the national interest. Some corrective forces to this free market in votes do exist. Some Congressmen have sufficiently high public profiles and/or safe seats that they can be confident of re-election on relatively small budgets, or on campaign support from the general public, rather than special interests. At the national level, Congressmen are held to account by organizations such as the League of Conservation Voters, which monitor and publish their voting records on environmental legislation. The influence of local interests, an important source of campaign funding, explains why the national Democratic and Republican political parties often exert little control over 'their' Congressmen, as demonstrated on the many occasions when Congressmen vote against bills proposed by a president of their own party.

The Senate and the House of Representatives are organized into major committees, which oversee departments or deal with broad issues, and multiple sub-committees. All legislation proceeds through one or more committees in both the House and Senate. Its progress is usually dependent upon the support of powerful committee, and sometimes sub-committee, chairmen who dictate the legislative agenda. Some environmental legislation must be dealt with by many different committees, whose jurisdictions over the environment often overlap. One source lists 13 House Committees and 9 Senate Committees with involvement in current environmental acts, including Public Works and Transportation, Appropriations and Energy and Commerce in the House, and Appropriations, Environment and Public Works, and Energy and Natural Resources in the Senate.[5]

Some of these committees, such as Environment and Public Works, are acknowledged to be 'distributive', or 'pork-barrel' committees, whose members are greatly influenced by a desire to direct any public expenditure towards their favourite projects or causes. Others are 'policy' committees, such as Energy and Commerce with its wide remit covering air pollution, health consumer protection and energy issues. The more intellectual, policy-oriented members are drawn to this committee, where difficult choices have to be made between competing areas. This fragmented legislative structure makes any coherent, comprehensive policy initiatives difficult to initiate or sustain, even without the inevitable interference from committee members responding to special interests.

Congressmen and their committees do not operate in a vacuum. A huge lobbying industry has grown up to influence their decision-making and assist them in drafting legislation. Environmental organizations and industry associations are increasingly involved in the details of the legislative process, often participating in informal 'issue networks'. These will typically include House or Senate Committee members, professional policy analysts from pressure groups and officials from the administration, such as the EPA.

---

[5] The Environmental and Energy Study Institute, *1993 Briefing Book on Environmental and Energy Legislation*, EESI, Washington, D.C., 1993.

The SO_2 Tradable Emission Permit scheme provides a good example of this process. Economists from the Environmental Defense Fund drafted many of the details of the scheme, in consultation with EPA officials and House Committee members. This scheme also had a novel feature which helped to ensure its success. It deflected the Congressmen who tended to concentrate on special interests from the overall impacts on the power industry (a source of campaign funds) by giving them state-by-state allocations of permits to fight over and hence an opportunity to boast of any gains they made over the draft proposals (see Chapter 10).

Once legislation has been enacted by the House, the Senate and the president, the executive branch must implement it. Since 1970, most major environmental acts have specified that the EPA should take on this task, by setting standards for environmental quality, promulgating regulations to protect health and the environment and funding environmental programmes. Even before Reagan's budget cuts in 1980, the EPA was overwhelmed by the amount and variety of tasks Congress laid before it. This mismatch between the ambitious legislation of Congress and the agency's resources has continued to this day. In the 1992 'Transition report' (an assessment conducted at the end of each four-year administration) the General Accounting Office (GAO) considered that the EPA needed a research agenda which would improve its risk assessment capabilities and provide better management of its existing data. The EPA maintained nine separate databases on pesticides but when a pesticide was spilled into the Sacramento river in 1991, the agency was unaware of information in its own files indicating that the pesticide caused birth defects. Other problems identified by the GAO are failure to meet statutory mandates, failure to establish priorities among programmes and failure to follow through planned improvements in management systems.[6]

In its defence, it must be pointed out that the EPA has been a political football for most of its existence and has lacked the stability to learn from its own experience. Beginning as a consciously created haven of environmental concern within the administration, it became a battleground between a

[6] United States General Accounting Office, *Environmental Protection Issues*, Transition Series, December 1992.

Democratic Congress which saw the environment as a 'wedge issue' with which to embarrass President Reagan who in turn regarded the EPA as an irritant standing in the way of wealth and prosperity. Recently there have been signs of a more strategic approach. This began under William Reilly's leadership, when the head of the EPA's office of air and radiation was encouraged to take some bold initiatives.

8.3.2. The Courts

America's fascination with legal processes exerts its baleful influence on environmental policy, as it does on so many other social issues. Since the early 1970s most environmental laws have given citizens legal rights to sue the administration for failing to implement legislative obligations. This arguably was a useful option when the Reagan administration was intent on frustrating the intentions of Congress and environmental groups sued the EPA for failing to meet mandates. However, it has just as often meant that environmental policy-making has fallen by default to judges, when the EPA has tried to implement vague primary legislation and subsequently been challenged by industry or environment groups.

This occurred more than once with respect to the Clean Air Act and its amendments. For example, the 1970 CAA was ambiguous as to whether areas which were in compliance with the national ambient air quality standards should nonetheless be subject to the same site-specific regulations on emissions which were applied to areas with poor air quality. The Sierra Club, one of the largest American environmental groups, sued the EPA on this point in 1972. The case was successful, and was upheld on appeal due to an even split of opinion in the Supreme Court, when one justice was absent.[7] This was truly a policy-making decision. In response, the EPA was compelled to develop the concepts of PSD review, aimed at 'preventing significant deterioration' in areas with good air quality, and 'offsets' in areas which did not meet national standards, to allow firms to offset emissions from new sources against reductions at existing sources. Congress later incorporated these approaches into the 1977 CAA.

---

[7] The Supreme Court is composed of nine justices, given lifetime appointments by the president of the day. Evenly divided decisions uphold the prior outcome in the lower courts.

Many objections have been raised to this legalistic approach to policy-making.[8] One concerns the competence of the judiciary. Judges are less well-informed on environmental and technical issues than the EPA or the committee members of Congress who specialize in these areas and draft much of the legislation. Faced with issues with which they are unfamiliar, judges attempt to make strict pedantic definitions of the original, often vague, legislation. Different judges arrive at differing interpretations, as illustrated by the above example, but so do the different Circuit Courts, each covering a specific region of the country. An extreme instance of this occurred when the Ninth Circuit agreed that the EPA could use external contractors to inspect industrial sites, while the Tenth and Sixth Circuits said that only EPA employees could make inspections. Overall, the influence of the courts has forced the EPA to act too strictly, following inflexible and often inappropriate procedures, and reduced the scope for negotiation between the EPA and industry on the most cost-effective methods of pollution control.

Congress is partly to blame for this situation, as it has often framed vague legislation in preference to making strict definitions of difficult concepts. The vagueness avoids alienating one or other interest group within Congress but allows multiple interpretations. In one case which came before the Supreme Court in 1980, Justice Rehnquist argued that Congress had delegated too much of its policy-making function to the executive (resulting in legal challenges to the EPA's interpretation of Congress's intent). One interpretation of Congress is penchant for excessively detailed rule-making on topics on which there is general agreement (for example the coke ovens legislation mentioned earlier) is that it emerges as a 'displacement' activity when the legislators are consciously, or subconsciously, avoiding areas of potential conflict over policy.

Recently, the costs of the litigation itself have begun to spin out of control. The main culprit is the Comprehensive Environmental Response, Compensation and Liability Act (CERCLA), commonly known as

[8] Perhaps the most damning indictment against the legalistic bias of environmental policy has been the charge of 'environmental racism'. This is the accusation that ethnic minority communities suffer disproportionately from pollution, one major reason for this being their lack of access to planning processes and especially to the courts, simply because they are poor.

'Superfund'. This 1980 legislation established a Superfund of $1.6 billion for the EPA to remove or clean up hazardous waste sites.[9] The EPA can require potentially responsible parties to clean up the sites or it can clean them up itself and then attempt to recover its costs from any party connected with owning or operating the site, or using it to dispose of waste under the act's strict, joint, several and (uniquely) *retroactive* liability.[10] This has resulted in an explosion of litigation. Parties identified by the EPA have attempted to dilute their liability by drawing in other parties (as the joint and several provision allows) or, in a limited number of cases, have sought to have themselves declared non-responsible. To date the legal costs far outweigh the costs of waste treatment, partly because treatment has been so badly delayed by legal wrangles. One site, contaminated by military production during the Second World War, has consumed legal fees of nearly $21 million since 1983 while nothing has been done by way of clean-up.[11] Estimates of the final cost of CERCLA range as high as $1 trillion, with legal fees accounting for most of this. At the beginning of 1991, 13 of the top 20 property and casualty insurance firms reported around 50,000 claims and 2,000 lawsuits – with typical legal costs of several million dollars each – already in the pipeline.

8.3.3. The States

The fifty US states have a greater or lesser degree of autonomy on environmental issues, depending on their willingness and capacity to fulfil the requirements of federal legislation. Some have been active and innovative, both in making local environmental policy and in implementing federal policies, while others have shown less interest and have been content to cede local control to the EPA.

---

[9] Congress seriously underestimated the number of hazardous waste sites around the country. The Superfund was reauthorized for $8.5 billion over five years, in 1986, and $5.1 billion over 3 years in 1990, effective from 1991.

[10] The joint and several provision means that any one responsible party can be held liable for all the clean-up costs, no matter how small their contribution to the pollution. Parties must sort out their share of responsibility among themselves. The retroactive provision means that parties can be held reponsible even where they were acting in accordance with all regulations in existence at the time the polluting activity occurred.

[11] James M. Strock, 'Wizards of Ooze', *Policy Review*, Winter 1994.

The Clean Air Act illustrates the relationship between the state and federal authorities. The 1970 CAA required states to develop State Implementation Plans (SIPs), to achieve national ambient air quality standards, as defined by the EPA, within a certain period of time. The SIPs can be a mixture of measures such as local industrial site licensing, traffic management, agreements with industry, building regulations or education. Most states have created their own local 'EPAs' to develop their environmental programmes in response to federal and state legislation. Some, however, have left environmental issues to existing agencies (e.g. the Department of Health in North Dakota), where they generally have a lower priority than elsewhere. If the EPA judges that a state's SIP is inadequate, it has the authority to develop and directly enforce measures designed to ensure the legislative timetables are met. The EPA has ten regional offices, each with responsibility for liaising with and, if necessary, directing, the environmental authorities in a group of states.

8.4. Death by a Thousand Rules

Any industrialist trying to make an honest buck in the US faces a daunting task, in relation to the confusing set of institutions and policy-makers he has to deal with in his efforts to make his voice heard in the legislative process. Where should he concentrate his limited financial resources? Lobbying Congress? Entertaining administration officials? Funding sympathetic 'environmental' organizations? Pressing ruinously expensive litigation? Can it really be worth the trouble? Perhaps it would be more rational to accept the inevitable and deal with the regulator when he comes knocking on the factory door. And yet, thousands of firms have decided they cannot bear the consequences of doing this, and willingly risk all in the courts and lobbies. Something about the day-to-day application of the law is giving them a very bumpy ride.

US environmental legislation is medium based. Separate environmental acts deal with air pollution, water pollution and solid waste, thus separately addressing the traditional environmental media of air, water and soil. Each act specifies, often in great detail, the actions which federal or state agencies must take to implement the objectives they lay down. In the past, the regulatory agencies have generally reacted by creating a new medium-based administrative division as each piece of enabling legislation has been enacted. In the EPA, for example, the air and water divisions are almost exclusively concerned with developing and enforcing the regulations flowing from 'their' legislation. This pattern is repeated at state level. The first major failing of the permitting, or licensing, system comes about, therefore, when overlapping permits are issued by regulators who are solely concerned with one medium, regardless of the overall impacts of their demands on other forms of pollution.

A second failing occurs through jurisdictional overlapping. One site may be regulated by federal EPA officials because it produces a pollutant which is locally above national ambient quality standards and for which the state has not developed an adequate strategy. Meanwhile, a state-wide body may be issuing permits relating to some other pollutants or activities, while responsibility for others has been delegated to city level, giving two more sets of regulators who are making demands on the site. 'Cities' in the US can be administrative bodies covering populations of only a few thousand, so a sister plant a few miles away may be dealing with different regulators, who are imposing different conditions and specifying different equipment for the same pollutant and process.

Inconsistent demands from Congress have also caused confusion. The Clean Air and Clean Water Acts of the early 1970s proved to be absurdly over-ambitious and naive. By the time of the 1977 CAA, Congress had to establish 'non-attainment' areas, where air quality standards were clearly not being met within the earlier time frames. Subsequent amendments have pushed some targets well into the next century.[12] The 1972 Water Act called for all discharges of pollutants into navigable water to be eliminated by 1985. This goal had to be revised only five years later, and again in 1987. Unfortunately, state and local permitting authorities pursued these moving targets with varying enthusiasm. Many companies were forced to take extremely costly emission reduction measures at short notice, to meet targets which were effectively repealed a few years later. Where local authorities pursued less stringent policies, no action was necessary. Animosity between industry and regulators grew apace.

---

[12] The Los Angeles–South Coast Air Basin is currently required to meet the national ground level ozone standard by 2010.

A prescriptive approach to environmental regulations has of itself been a source of frustration and excessive costs and has stifled new technologies and processes. When it promulgates a new regulation, the EPA produces guidance on the technologies which can meet the desired emission standard. The standards themselves can be based on maximum achievable control technology (MACT), reasonably achievable control technology (RACT) or some other formulation, depending on the enabling legislation. The EPA uses various methodologies for determining what is 'reasonable' or 'maximum'. In the area of toxic air pollutants, for example, the performance level of the top 12% of existing plant is used to set the standard. Information on newly approved technologies should be disseminated to all regional EPA, state and local permitting authorities, but this process is – according to EPA officials – very slow and unreliable. Lack of continuous centrally approved updates, and fear of litigation, makes permit writers very conservative. They are unwilling to authorize new technologies and processes, so the EPA guidance becomes a *de facto* requirement and technological lock-in occurs.

The manner in which officials deal with technical risk also reduces the experimentation which is required for new technologies to be successfully demonstrated and enter the marketplace. Regulators will often allow novel equipment to be installed only if additional, proven control measures are introduced further downstream in the production or treatment process. Most will also insist that a novel technique which falls only slightly short of the required target is removed and replaced with traditional control technology. The Department of Commerce and the EPA recognize that these attitudes are a significant disincentive and a threat to the US environmental technology industry. Many US manufacturers have sought demonstration sites in Europe or Japan, where regulators are perceived to be more tolerant of new technologies. These attitudes compound the negative effects of a fragmented market, represented by the patchwork of regulators, unpredictability of regulatory requirements, caused by the slow gestation of primary legislation, about-faces in Congress and the courts and political interference with the EPA, and a single-medium focus which is biased against the all-round benefits typical of cleaner production technologies.

Finally, the whole regulatory effort is, quite simply, badly run. Some of the General Accounting Office's criticisms of the EPA were outlined in

Section 8.3.1. The GAO also drew attention to the inadequacy of efforts by EPA headquarters to ensure that all EPA regions and states enforced drinking water standards, and failure by some regions to adhere to guidelines for assessing penalties for non-compliance with regulations. This lack of central authority and control causes further confusion and fragmentation in the regulatory process.

The perilous state of the industrial licensing system, i.e. the practical application of Congressional legislation and the EPA's regulations, is a major reason for the enormous costs borne by US industry in complying with environmental legislation. Estimates of these costs vary wildly, come from numerous interested parties and are usually terrifying, with figures of hundreds of billions of dollars commonly bandied about. These can be misleading because of the enormous size of the US economy, where even taco chip consumption runs into billions of dollars. However, the scale of the problem is summed up by the EPA. Surveying current and pending legislation in 1990, the agency estimated that without legislative and regulatory reform the annual cost of pollution control could reach 2.8% of GDP by the year 2000.[13]

8.5. Evolution I: California Dreaming?

California's environmental agenda is dominated by the unique problems generated by the unfortunate conjunction of the Los Angeles metropolis with the South Coast Air Basin. The air basin is formed by a ring of mountains which hem the city in against the Pacific coast. A meteorological phenomenon known as a temperature inversion occurs when a layer of cold air moves inshore from the ocean and becomes trapped against the mountains. This static layer of air over greater Los Angeles extends from ground level to around 1,000 metres.

Some simplifications help to illustrate the human dimensions of the air quality problems of Los Angeles. When a temperature inversion occurs, each of the 15 million Angelenos is effectively living in a sealed cube measuring around 150 metres on each side. If a typical inhabitant drives for 2 hours

[13] Quoted in Clement B. Malin, 'Politics, Economics and Environment: Experience of the US Oil and Gas Companies', in N. Steen, see note 2.

each day in a car with an engine capacity of 2 litres, his entire supply of air will have passed through his engine in around 100 days. After that point he may as well breathe air straight from the exhaust.[14] In reality things never get this bad, since temperature inversions tend to be daily phenomena, not all Angelenos drive, and the 'box' is not perfectly sealed as there is some exchange with the overlying and surrounding air masses. On the other hand, there are many other sources of air pollution, including heavy industry and residential and commercial activities. In 1992, vehicles accounted for three-quarters of nitrogen oxide emissions and only half of the emissions of hydrocarbons.[15]

Prompted by the problems in Los Angeles, California has vigorously tackled pollution. As a result, peak levels of ozone in Los Angeles, which regulators consider to be a good indicator of general air quality, dropped from 680 parts per billion in 1955 to 300 per billion in 1992, despite a tripling in population.[16] In achieving this, California has developed a regulatory system which displays all the costly and frustrating features – such as fragmentation, disincentives to innovation and poor prioritization – identified in the previous section. Criticism by businesses of the difficulties of negotiating the system reached a peak during the recession of the early 1990s. This encouraged the state governor, Pete Wilson, to begin a radical overhaul of Californian environmental policy, allied with encouragement of new environmental technologies, as part of a wider strategy to replace high technology jobs which were being lost in the defence industries.

First, and most radically, the governor plans to replace the patchwork legislative framework inherited from Congress (separate air, water, waste acts, etc.) with a unified environmental statute. Officials at the Californian EPA (Cal/EPA) see this as an essential part of a process of evolution from a medium-based approach to an industry sector approach.

Initiatives which are already under way are designed to tackle problems at the operational level. Enforcement teams, composed of inspectors with

[14] Readers are advised *not* to try this.

[15] J. M. Lents and W. J. Kelly, 'Clearing the Air in Los Angeles', *Scientific American*, October 1993.

[16] This still leaves Los Angeles with the worst 'non-attainment' ratings in the country, for ozone, carbon monoxide and fine particulates.

combined experience of all media, are visiting industrial sites and making comprehensive (rather than contradictory) recommendations. Coordination of state, regional, local and selected federal agencies is resulting in designation of lead agencies for multi-media permitting. Cal/EPA is promoting harmonized state-wide standards: producing a best available control technology guide for air pollution, modifying operations in the nine regional Cal/EPA boards and unifying six separate state and local hazardous waste programmes.

In addition to reducing the administrative difficulties for businesses, these moves towards consistent state-wide standards will encourage technology development. Other measures which are designed directly to promote novel technologies are under way or planned. The California Environmental Technology Partnership (an association of Cal/EPA, the California Trade and Commerce Agency and the Environmental Technology Advisory Council) has published a strategy for encouraging environmental technologies to be developed in California.[17] A new body will certify methodologies for risk assessment of new pollution control devices, protective equipment, or cleaner technologies, processes and procedures. A formal certification programme will use independent, third-party evaluations to certify that new technologies meet Californian regulations. Initially this programme will apply to hazardous waste technologies and will standardize testing procedures, certify and publish test results, acknowledge that adequate demonstration has occurred and identify those legally designated standards, such as reasonably available control technology (RACT), which have been achieved. To provide technology developers with prospects of a reasonable market, the programme may designate a minimum 'lifetime', e.g. five years, during which state permitting authorities should accept a technology which has been certified for a specific purpose.

These initiatives, combined with joint industry–agency research and demonstration projects, are intended to 'establish California as the best place in the world to test and demonstrate environmental technologies'.[18]

---

[17] California Environmental Technology Partnership, '1994 Strategic Plan for Promoting California's Environmental Technology Industry', Department of Toxic Substances Control, California Environmental Protection Agency, January 1994.
[18] Ibid.

Underlying the strategic plan is an explicit desire to increase California's share of national and international markets for environmental technology. Export promotion activities include foreign market guides and annual trade missions to Japan and Europe.

Other state agencies are forging new kinds of relationships with regulated industries. The South Coast Air Quality Management District (SCAQMD), responsible for non-vehicle emissions in the Los Angeles area, is leading the way (and also developed many of the permit streamlining processes mentioned earlier). SCAQMD derives most of its income from the fees it charges for permits and penalties raised from non-complying businesses. In a new 'softer' approach, its experts have been visiting firms which admit to difficulties in meeting regulations, to give technical advice and assistance. Penalty proceedings are waived if the firm agrees in advance to follow the SCAQMD recommendations.

Increased consultation on permitting processes has led to simplified procedures. A business with many identical pieces of equipment used to pay a permit fee on each one. Now SCAQMD requires a permit only for the first example of any type. Ironically, these improvements have sharply reduced SCAQMD's income and precipitated reductions in staff. The institutional funding arrangements for SCAQMD and other agencies dependent on fees will need to be overhauled, if progress towards further industry–regulator cooperation, and away from penalties and excessive permit-writing, is to be maintained.

SCAQMD has also pioneered a novel market-based regulatory instrument: the Regional Clean Air Incentives Market, or RECLAIM. Companies emitting more than four tonnes per year of NOx or SO_2 from equipment which is usually subject to permitting procedures can choose to participate. (There are plans to add VOCs in the near future.) Participants are freed of all permit requirements (and fees) and are issued annually with credits for their emissions. Firms can trade their credits with one another, allowing some to reduce their emissions and sell credits, and others to avoid expensive reduction options by purchasing credits. Each year after the start of the scheme in 1994, SCAQMD will sharply reduce the total number of credits available, reducing NOx by 75% and SO_2 by 60%, by 2003. In an imaginative twist, firms can gain credits by buying, and then scrapping, 'clunkers', i.e. old, polluting cars.

Not all of these policy and regulatory innovations are secure. The governor's plan for a unified environmental statute may be difficult to reconcile legally with federal legislation. Federal powers impinge in other ways. A current lawsuit contends that California's existing State Implementation Programme will fail to meet the deadlines for attaining national air quality standards set out in the 1990 CAA. The lawsuit seeks to force the federal EPA to use its powers directly to impose and enforce regulations in non-compliant states. The RECLAIM programme reduces the danger of the lawsuit succeeding, but the state may yet have to accept an enhanced vehicle inspection and maintenance programme, which the governor has been resisting on cost grounds.

Attitudes to permits and penalties must evolve, to ensure that the cooperative approach pioneered by SCAQMD survives. Environmental groups and federal legislators are used to regarding high rates of permitting and penalties as an indication of a successful regulatory agency. Some pressure groups have been quick to criticize the drop in penalties and permits at SCAQMD, even though more firms than ever are complying with regulations. Legislators must urgently address the need for new funding arrangements for regulatory agencies.

8.6. Evolution II: America the Brave?

While several states, such as California, have begun to tackle the day-to-day complexities and failings of the regulatory system, the national stance on environmental issues – as promoted by the federal government – has changed beyond recognition since the Reagan presidency. The Clinton administration, building on the thawing of relations between President Bush and Congress, is pressing ahead with legislative and regulatory reforms and encouraging active, voluntary involvement by industry in environmental programmes.

The most public change in American environmental policy has been the Clinton administration's commitment to stabilizing greenhouse gas emissions at 1990 levels by the year 2000.[19] At the Earth Summit in Rio in 1992, William Reilly, President Bush's progressive EPA Administrator, was keen for America to make this commitment. But Bush – a naturalized Texas oil man

[19] W.J. Clinton and A. Gore, 'The Climate Change Action Plan', October 1993.

– baulked at embracing the spirit of the Earth Summit, with all its implications for energy-profligate America, and was widely condemned as a result. The Clinton Climate Change Action Plan recognizes that stabilizing emissions could have widespread impacts on society and the economy. It establishes new, voluntary programmes with industry and business, both to allay the fears of these groups and out of a conviction that cooperation, rather than regulatory decree, is the best way to proceed.

The most important voluntary initiative is the 'Climate Challenge' programme. The Clinton team approached the country's electric utilities, through their industry association, the Edison Electric Institute, to discuss how they could contribute to national reductions in CO_2 emissions. By spring 1994, 80 utilities, representing 90% of the country's electricity consumers, had signed letters of intent to enter into formal agreements with the Department of Energy at a later date. In the agreements, each company will be expected to commit itself to an emission reduction strategy, based on its own choice of measures from a range developed by the Edison Electric Institute and its member companies. These voluntary reductions will be recorded by the Department of Energy, under a provision of the 1992 Energy Policy Act, which foresaw, but fell short of mandating, a possible future need to reduce CO_2 emissions.[20] The Climate Challenge programme may well transform that provision into a self-fulfilling prophesy.

One potential difficulty which has been recognized by the Edison Electric Institute, but not yet tackled by the administration, is how to cope with free-riders. Many, mostly small, utilities have given no indication that they will participate in an agreement. At the same time, the US electricity market is becoming more competitive, as a result of the 1992 Energy Policy Act breaking down exclusive regional monopolies. When the cost-free voluntary measures have been exhausted, participating utilities considering further emission reductions will inevitably face the possibility of losing customers to non-participants who are able to offer cheaper electricity rates. At this point, some kind of 'back-marker' regulatory intervention by the administration will be required to save the agreements. The EPA would

[20] The Energy Act also requires the EPA to record efforts which utilities make to offset their emissions elsewhere, e.g. through tree-planting programmes in the United States or abroad.

already be in a strong position to introduce some kind of market-based regulation, thanks to the existence of its register of voluntary reductions and offsets. It could choose a CO_2 emission trading scheme, with registered emission reductions being transformed into emission permits. Alternatively, the reductions on the register would translate easily into tax rebates, if a CO_2 tax were introduced.

Another voluntary scheme, the 'Climate-Wise Companies' programme, involves the Department of Energy and EPA establishing energy efficiency goals for industries, and allowing individual firms to choose the most cost-effective of a range of measures, including industrial process changes and raw material and fuel substitutions. As with the Climate Challenge programme, emission reductions and offsets will be registered on the database established by the Energy Policy Act. The same difficulties over competitive pressures from non-participants will undoubtedly show up sooner or later.

One of the prime motives behind the voluntary approach of the Climate Change Action Plan and its focus on 'partnership' with business, is the administration's desire to avoid the legalistic quagmires of Congress and the courts. In the early 1990s, a peculiar American variant on voluntary agreements emerged, as a result of the administration's desperation to avoid legal challenges to the EPA's regulations. This is known as regulatory negotiation (reg-neg). Reg-neg consists of the participation of all interested parties (i.e. businesses, national environmental groups, local action groups, etc.) in the formulation of a regulation. The hope is that they will reach agreement on the detailed form of the regulation and so refrain from challenging it in court.

Unfortunately, reg-neg does not seem to have encouraged business or environmental groups to abandon the over-legalistic mindset with which they approach regulations. It may even have introduced feelings of broken trust and bitterness which were not present when the two sides automatically distrusted one another. For example, in early 1994 the EPA announced its plans for implementing a regulation on vehicle fuel additives. All sides had thought that they understood the scope of the proposed regulation, and were content with it. When the refining companies saw the EPA's plan for implementing the regulation, they were furious at provisions which their lawyers interpreted as being outside the remit of the regulation. Normally they would have sued the EPA, and had judgment in court, but most of the companies felt bound by the reg-neg process. Instead, they muttered dark warnings about never cooperating with the EPA again.

The administration will need to nurture its first attempts at true voluntary agreements with the care that the Dutch, for example, have taken, if they are to avoid irretrievably tarnishing the reputation of agreements within the business community. (Most environmental groups start out from a position of hostility to voluntary agreements and will only grow to accept them if they clearly deliver the goods.)

Meanwhile, the small revolution in attitudes at the EPA which began with William Reilly and the policy innovations within the Office of Air and Radiation, especially during the development of the acid rain strategy in the 1990 CAA, shows signs of spreading and becoming more permanent under the new administrator, Carol Browner. A technology innovation strategy is under discussion. A draft outlines the problems with the existing regulatory system – technology lock-in, fragmented market, single-medium permitting and regulatory uncertainty – and proposes many of the solutions favoured in California.[21] In addition, the EPA proposes 'soft-landings' if innovative technologies narrowly fail to meet permit requirements.

The EPA's proposed technology innovation strategy includes measures to promote US environmental technologies abroad. One such measure is the US Environmental Training Institute (USETI) – a public–private partnership launched in 1991 to build environmental institutions and capacity in industrializing countries. The discussion document suggests that USETI's training courses on risk management, pollution prevention and environmental management will stimulate demand for US technology and expertise. This confirms a recent trend in the administration towards identifying opportunities where export of regulatory expertise and advice can be a precursor to export of technology. In another case, Japanese environment officials have reported American pressure for Japan to bring its standards for hazardous waste and contaminated land into line with those in the US. This can be interpreted as an attempt to recover some of the absurd costs of the Superfund and its attendant legal wrangles through exports of remediation technologies.

---

[21] EPA, 'Technology Innovation Strategy: External Discussion Draft', EPA, January 1994. Available on Internet: telnet to ebb.stat-usa.gov, filename: TECSTRAT.EPA.

In an attempt to break from its medium-based traditions, the EPA has launched a pilot programme under its 'Common Sense Initiative', which will focus regulatory activity on individual industrial sectors. It has selected six industries – automobiles, electronics and computers, iron and steel, metal plating and finishing, oil refining and printing – for its 'Green Sectors' programme, after gauging the degree of industry interest, willingness to commit resources, and opportunities for significant pollution prevention and risk reduction. The selection is intended to represent a diverse range of industries.

While these individual initiatives have been pressing ahead, a more fundamental review of the entire US environmental protection system has been under way. According to the White House Office of Environmental Policy, their approach to date has been to step back from the existing situation, develop a comprehensive, cohesive view and then propose widespread reform. This will encompass flexibility of regulations, ecosystem management and better coordination of agencies at federal, state and local levels. A major restatement of the US framework for environmental policy will result from this exercise, 'in 1995, at the earliest'. Economic and social issues, such as 'environmental justice' will feature heavily in the new policy.

In this connection, the President's Council on Sustainable Development (PCSD) will make explicit, and popularize, the link between regulatory reform and sustainable development.[22] White House staff feel that industry now realizes it is better to have a seat at the table. In their view, the PCSD provides this and is a model of how to build trust between industry and government. The PCSD is organized into several issue-specific task groups and sub-groups. The 'Eco-efficient manufacturing' sub-group is exploring the potential for pollution prevention, design for the environment and a 'cradle-to-cradle', i.e. closed cycles, approach. White House staff are interested in new approaches elsewhere, such as the Dutch plan for a sustainable economy. They are convinced of the benefits of such an approach, in terms of setting a cohesive, long-term agenda, thus reducing the uncertainty

[22] The PCSD has 25 members, including the EPA administrator, Commerce Secretary, Interior Secretary, heads of industrial firms and directors of environmental groups. Its task groups cover sustainable communities; eco-efficiency; population and consumption; climate change; and sustainable agriculture.

for industry, and hopeful that something similar could be implemented in the US, despite a very different economic structure.

8.7. Conclusions

The tragedy revealed by examining US environmental policy is that the unnecessary cost and inefficiency about which US business complains so noisily has come about largely by accident. Congressmen, bureaucrats and, often, judges have been more ignorant of industry than indifferent or hostile to it. The 'burden' of environmental regulations, an idea which Ronald Reagan and Margaret Thatcher accepted without question, could be more correctly described as the burden of the US approach to environmental regulations. In other countries, where an informed and continuous dialogue between regulators and industry is possible, flexible, evolving approaches allow innovation to take its course.

The United States faces prodigious challenges in unwinding the polarized, damaging approach to formulating and implementing environmental policy which it has created for itself. The excessive costs imposed by its rigid, legalistic approach have rightly prompted a major reappraisal. However, powerful conservative trends may stifle and even roll back the progress already made towards more flexible industry–regulator relationships. Chief among these is the prevailing US mindset on all social issues, which seeks to frame problems in rigid, black and white legal terms. Environmental issues, which are often fluid and uncertain and where compromise and cooperation would be sensible, are instead forced into endless rounds of judgment and appeal. An enormous and influential cadre of legal professionals gives this process an almost irresistible momentum. If the 'agency capture' feared by Congress in the 1970s has come about, it has been perpetrated by lawyers, not industrialists.

Congress is itself a captive of lawyer-lobbyists, who facilitate an increasing throughput of ever more detailed and constraining legislation. Pork-barrel references to specific spending projects reinforce the rigidities. It is difficult to imagine that a unified environmental bill, setting out only broad objectives, principles and processes, as envisaged by the White House, could be passed by legislators without Congress first undergoing a wholesale reform.

Part 3

Chapter 9

Driving Technology

9.1. Introduction

Everyone believes in the right to own a big, beautiful car. Attempts by officials to interfere in driving habits are generally resented. Officialdom is, by and large, indulgent of drivers: in most large industrial countries, governments view car sales as an important indicator of the health of the economy and, in those countries with their own car industry, politicians and industrialists agonize over every movement in the ratio of car imports to exports. At the same time, because of its very popularity, the car is one of the most pernicious sources of health and environmental problems faced by modern societies. As this became more and more apparent in fume-choked Western cities, vehicle pollution developed into the defining environmental issue of the late 1960s and early 1970s: the clash of wealth and consumption with health and the quality of life.

Demands to improve urban air quality, combined with government obsession over the car industry's status as a bellwether of the national economy, led to a linkage between emission standards and competitiveness in the mid-1970s. In some countries this linkage had more credence among politicians and industrialists than in others. This chapter explores how beliefs about the commercial significance of emission standards have been influenced by the internal organization and innovative capacities of the major car firms and by the relationships between national vehicle industries and policy-makers. Tracing the history of American, Japanese and European experience up to the present day reveals just how badly the rest of the world has missed the point about Japan's success. Japanese car manufacturers have succeeded through sheer commercial and especially manufacturing prowess. Behind this has been their development of the lean production system which has given them such an advantage over mass manufacturing firms that other factors in their commercial environment – such as recessions and environmental policies and regulations – have little bearing on their long-run success.

California's requirement for electric vehicles to be supplied in large numbers from 1998 is the latest attempt to link emission standards and competitiveness. It serves only to illustrate that while policy-makers may be able to force firms to commercialize technologies, no one can force technologies to appear in the first place.

9.2. The Nature of Vehicle Pollution

For policy-makers the major concern relating to automobile pollution is its impact on local air quality and the consequent health effects, primarily in urban areas.

The Industrial Revolution resulted in poor air quality and often severe impacts on health through uncontrolled emissions of particulates, sulphur dioxide and heavy metals. By the middle of the twentieth century these problems had become so acute in Western Europe, the United States and Japan that governments instituted controls on industrial pollution. These controls coincided with shifts in industrial structures, with the net effect that industrial pollution sources now make relatively a much smaller contribution to the unhealthy air found in urban areas within the developed countries.

However, just as the direct legacy of the Industrial Revolution has been successfully addressed, so the automobile – the most potent symbol of the Consumer Revolution – has risen to take its place as the most important contributor to urban air quality problems.

Vehicles cause harmful emissions in a variety of ways. Most important of these are tailpipe emissions, i.e. harmful exhaust gases. However, evaporative emissions, e.g. at fuelling stations during filling and from vehicle fuel tanks, are also significant and are increasingly being regulated. The box listing the pollutants produced by vehicles is not exhaustive, but covers the major offenders.

The environmental impacts of vehicle pollution are diverse. They range in scale from the truly local, such as roadside exposure to carbon monoxide, to the regional and global impacts of acid rain and the global impacts of carbon dioxide. The chemical behaviour of vehicle pollution is also complex. As

Major pollutants produced by gasoline and diesel vehicles

- *Hydrocarbons*

The hydrocarbon fuel which may evaporate before reaching the engine of a vehicle will also turn up in the tailpipe emissions if combustion in the engine is incomplete. Hydrocarbon emissions are a major component of gasoline vehicle emissions. However, diesel engines operate with an excess of air which generally ensures more complete combustion of the fuel, and lower levels of hydrocarbons in diesel exhausts. Volatile hydrocarbons from gasoline vehicles are a major component of the class of pollutants generally referred to as VOCs, or volatile organic compounds. These compounds react to produce ozone, a secondary pollutant which is highly reactive, attacking the lungs and respiratory system and stinging the eyes, at the concentrations typical of urban areas with poor air quality. Ozone causes short-term respiratory difficulty and longer-term exposure reduces lung function and retards lung development in children. Some of these volatile hydrocarbons are also toxic in their own right. Benzene and 1,3-butadiene are genotoxins, known to cause cancer, at very high concentrations, in toxicological tests. However, epidemiological evidence suggests they have a very minor effect at the low concentrations produced by vehicles. For example, benzene occurs in cigarette smoke at concentrations several orders of magnitude higher than the levels encountered in even the most polluted cities. Even if the majority of cancers caused by smoking are attributed to benzene, whereas in reality a cocktail of carcinogens is involved, this puts a very low ceiling on the possible effects of benzene from vehicles. Benzene also occurs naturally in pine forests at higher levels than are encountered in cities, and epidemiological evidence from forest dwellers again puts limits on its possible effects.

- *Particulates*

When diesel fuel hits the internal walls of the combustion chamber, a proportion is quenched and fails to ignite. This forms very fine carbon particles which are emitted as particulates – soot or 'diesel smoke'. Particulates small enough to remain suspended in the air (as suspended particulate matter, or SPM) encompass a large range of particle sizes which have different effects. Most worrying are very fine particles of 10 microns diameter and less (PM-10). These are inhaled deep into the alveoli of the lungs, and are suspected of contributing to lung diseases such as asthma, bronchitis and emphysema. A cocktail of organic and inorganic compounds is associated with particulates, which are suspected of contributing to lung cancer.

- *Carbon monoxide*

Carbon monoxide is a poisonous gas which has debilitating effects, such as headaches and fatigue, at sub-lethal levels and contributes to respiratory and heart diseases after prolonged exposure at low levels. As with tailpipe emissions of hydrocarbons, carbon monoxide is associated with low levels of air (i.e. oxygen) in the engine during combustion. In this case there is insufficient air for complete combustion to carbon dioxide.

- *Nitrogen oxides*

Nitrogen oxide (NOx) formation in vehicles is fundamentally the same process as thermal NOx formation in coal-fired power stations (see Chapter 10). Indeed, NOx will be produced whenever high-temperature combustion occurs in air with an excess of oxygen. Diesel engines, which operate ideally with an excess of air, are especially prone to high levels of NOx formation, although gasoline vehicles also emit significant quantities. NOx has direct effects on respiration, particularly among the very young, the elderly and people with existing respiratory problems such as asthma, reducing lung functioning through mechanisms which are not yet fully understood. It is also, along with hydrocarbons, a precursor to the formation of ground level ozone and, as discussed in Chapter 10, contributes to acid rain.

well as the primary pollutants emitted by vehicles, the secondary products and the conditions under which these are most likely to be produced must be taken into account. For example, ground level ozone formation is affected by climate and local geography: the Los Angeles area is surrounded by mountains which trap polluted air and allow time for the abundant sunlight to promote the chemical reactions which lead to ozone formation.

This complexity has provoked a range of responses among regulators and industry. These responses can be characterized as being *fuel-based*, *vehicle-based* or *behavioural*. At times, the responses can overlap or interact. For example, catalytic converters require unleaded gasoline, and efforts to reduce driving speeds and 'aggressiveness' might produce limits on octane values for fuel and on engine power. Reducing traffic levels in cities by providing better public transport or banning cars from city centres might allow private cars to have higher emission levels, with no impact on health. However, this study focuses on a series of vehicle-based technological innovations, where interactions with fuel quality and behavioural changes are minor, or straightforward.

9.3. Vehicle-based Technology Responses

Technological responses to vehicle pollution fall into three broad categories: optimizing combustion conditions in the engine, fitting emissions clean-up devices or switching to cleaner fuels.

Optimizing combustion conditions will only get you so far, and typically it involves a trade-off between two or more pollutants. For example, burning gasoline with low levels of air suppresses NOx formation but increases emissions of unburnt hydrocarbons and carbon monoxide. Increasing the proportion of air cuts emissions of hydrocarbons and carbon monoxide but increases NOx emissions up to a certain point, where NOx formation will begin to fall once again. To complicate the situation further, there is an ideal air/fuel ratio, known as the stoichiometric ratio, where maximum energy is derived from the fuel.

Measures to alter combustion conditions include retarding the emission timing, electronic fuel injection, chamber redesign and electronic engine management. These and other measures in combination with high compression ratios allow gasoline engines to operate with much increased proportions of air. This development path has led to the *lean-burn* engine which operates with an air to fuel ratio of up to 22:1, compared with 15:1 for a conventional engine. Hydrocarbons, carbon monoxide and NOx are all substantially reduced, in comparison with a conventional engine.

When modifications to the combustion conditions fail to produce sufficient emission reductions, clean-up devices can be fitted to exhausts to aid the process. In gasoline vehicles, *catalytic converters* have been fitted to exhaust systems for many years in some countries. Catalytic converters contain a catalyst – usually a metal such as platinum or rhodium – which promotes the conversion of the pollutants to harmless gases. Catalytic converters come in two main varieties: the oxidation catalyst and the three-way catalyst. Oxidation catalysts became widely used in the United States and Japan in the 1970s. They convert hydrocarbons and carbon monoxide to water and CO_2. From the late 1970s these countries began to adopt the three-way catalyst, which has the additional ability to convert NOx to harmless nitrogen gas.

The lean-burn approach did not develop beyond the early prototype stage until the 1980s. When the European Community finally got serious about auto emission standards, the political and industrial champions of catalytic converters and lean-burn engines engaged in a strategic battle over regulations which is widely cited as an example of the importance of environmental standards for national competitiveness. This story is traced in more detail later in this chapter.

Catalytic converters are now expected to attain emissions levels which would have been unimaginable only five years ago. For some, however, even this is not enough. California is seeking to ensure that an ever-increasing proportion of its vehicles will produce zero emissions. Only a completely new energy source, e.g. electric batteries, can fulfil this goal. This is the most extreme example of what can be achieved by switching fuel, a strategy which also includes the use of natural gas or ethanol derived from agricultural products. California's move towards electric vehicles is shaping up as the next major strategic battlefield for the world vehicles industry. For reasons discussed later, this initiative may have consequences as unintended as its origins were confused.

9.4. The NOx Controversy in the United States and Japan

9.4.1. History of NOx Regulation

Vehicle emission regulations were first applied in the 1960s, in Japan and in California. These early emission limits, such as the 1966 ruling in Japan limiting carbon monoxide emissions to 60% of the previously uncontrolled level, were relatively uncontentious. These regulations were based on emission levels which were already being easily achieved in some production vehicles. Their net effect was primarily to enforce good engine/exhaust design among manufacturers. The real issues of conflict between manufacturers and governments and between one government and another did not surface until later. At the root of these conflicts was a disagreement over the state of advancement of the technologies, and how far regulations could be expected to push their development.

In the United States, a section of the 1970 Clean Air Act known as the 'Muskie Law', after the Senator who proposed it, required a 90% cut in carbon monoxide and hydrocarbons in exhaust emissions, relative to their historic uncontrolled levels, by 1975. This was based on assumptions about likely improvements in the newly developed oxidation catalyst. The Muskie Law also called for large cuts in NOx emissions.

Japan had been gradually reducing allowable CO and hydrocarbon (HC) emission levels, and had also been setting standards for NOx emissions. Taking its cue from the American legislation, the Japanese Environment Agency also called for a new round of controls on emissions. Japanese manufacturers subsequently testified to the agency that they could achieve 5% reductions of HC and CO and a 70% reduction of NOx, by 1975, using a combination of an oxidation catalyst and a design change known as exhaust gas recirculation (EGR).

The agency caused controversy when it sought even tighter controls on NOx. The major manufacturers protested that its proposal for a 92% cut in NOx by 1976 would be technically impossible. The Industrial Bank of Japan produced a study suggesting that higher prices, increased fuel consumption and poorer performance would result, hitting sales and, ultimately, losing over 90,000 jobs. A government advisory body, the Central Council for Environmental Pollution Control, was instructed to examine the feasibility of the proposed target, which would impose a fleet-average limit of 0.25 grams of NOx per km.

Meanwhile, responding to strong public concern over local air quality, seven cities commissioned their own team of experts to study the issue. The team drew heavily on evidence from Honda and Mazda, both minor league manufacturers at that time. These smaller, more vibrant firms had pursued the development of the novel three-way catalytic converter – a means of operating a catalyst so that CO, HC and NOx are converted simultaneously to harmless gases – and were more than happy to share their information with the inquiry team.

The inquiry concluded that auto manufacturers already had the technological expertise to meet the standard and that imposing the standard would ensure final development and implementation of the technology. This conclusion was accepted at national level, although the deadline for implementation was extended to take effect in the 1978 model year. Manufacturers were offered tax breaks for meeting the standards a year early, which many did, even as their low-NOx vehicles set new records for fuel economy.[1]

Back in the United States, the big three car manufacturers had been lobbying the US Environmental Protection Agency, which was responsible for turning the Muskie Law into workable regulations. Their protestations that the proposed NOx standards would be impossible to achieve led to a study by the National Academy of Sciences, published in 1974. This formed the model for the Industrial Bank of Japan's gloomy report, and concluded that the Muskie Law would be technologically difficult to implement and would damage the US car industry. (This was despite the fact that Honda and Mazda were already demonstrating prototype cars in Japan which met the Muskie Law standards. Indeed, Senator Muskie rode in one when he visited Japan in 1973.) The study activated a postponement clause, putting off the NOx deadlines until 1983.[2]

---

[1] U. Kazuhiro, 'The Lessons of Japan's Environmental Policy: An Economist's Viewpoint', *Japan Review of International Affairs*, 7, 1, 1993, p. 30.

[2] The 1970 Clean Air Act resulted in a NOx limit of 1.9 g/km (grams per kilometre) in 1974 and called for a limit of 0.6 g/km. A compromise limit of 1.2 g/km, achievable with oxidation catalysts, was introduced in 1977. It was not until 1983 that the EPA approved a limit of 0.6g/km (*Transport and the Environment*, OECD, Paris, 1988, p. 89).

9.4.2. A Dirty Trade War

Having persuaded US officials that the Muskie Law was not technologically achievable, the US auto manufacturers petitioned the State Department with their view that the Japanese NOx standards represented an unfair trade barrier. The US government chose to interpret the Japanese standards as a non-tariff import barrier, perhaps unaware of the grass-roots public pressure which had forced the Japanese government into legislation. Under threat of a trade war, Japan agreed to exempt imports from the standard for a further three years. When the regulation was finally adopted in Japan, it gave three separate compliance dates: 1 April 1978 for new model cars, 1 April 1979 for all new cars and 1 April 1981 for imported cars.[3]

9.4.3. Contrasting Policy Styles

Several lessons can be drawn from this. First, an important point to bear in mind is that the proposed emission controls arose from strong public concern over air quality in cities. In the United States, the link with this constituency was effectively severed when the regulatory process came under the control of the EPA, after enactment of the Clean Air Act by Congress. In Japan, by contrast, local anxiety found expression in the initiative by the group of metropolitan authorities which ultimately salvaged the Environment Agency's proposal for a strict NOx regulation. This is a clear example of how the responsiveness of Japanese regulatory processes to public concern contributes to the quality of decision-making (see Chapter 7). Local concerns found political expression throughout the formulation of the policy and the final regulation, flushing out relevant technical information which would not otherwise have come to the fore.

The attitudes of the vehicle manufacturers in each country were also important to the regulatory outcome. Japanese manufacturers willingly accepted a system whereby they shared their technical expertise with the regulators. During the NOx debate, the fledgling Japanese Environment

---

[3] Japan Ministry of Transport, Ordinance No. 47, 22 December 1977, cited *in Motor Vehicle Pollution Control in Japan* (4th Revision), Japan Environment Agency, August 1993. It is ironic that the United States now worries about unfair competition from developing countries with less stringent environmental regulations on manufacturing *processes*, when in the 1970s it forced Japan to accept US *products* which increased pollution.

Agency sought evidence from only the major manufacturers, who were genuinely unsure of how the standards could be met. As we have seen, this oversight was corrected by the action of the metropolitan authorities. However, the major manufacturers responded by launching their own development programmes, rather than trying to discredit the information from Honda and Mazda. A more formal system for gathering evidence, from industry generally, has now evolved. This is based on a willingness on the part of firms to share their information with the government. This system is discussed at greater length in Chapter 7.

9.4.4. Contrasting Industry Attitudes to Innovation Policy

In the United States, the vehicle manufacturers were implacably opposed to the Muskie Law. The National Academy of Sciences study was based on evidence from the big three manufacturers which suggested that the technologies were immature and would be too expensive for the industry. It is difficult to believe that this was the real state of knowledge within Ford, GM and Chrysler, given that Japanese companies were already successfully demonstrating these technologies.

A difference in attitude towards working with government is only one element contributing to this obstructiveness. Differences in industrial structure also played a part. The structure of the industry in the United States, with three large dominant firms, was much less competitive than in Japan, where nimble new entrants were pressing hard on the older firms such as Toyota and Nissan. Japanese firms were engaged domestically in a furious technological battle across all fronts, with emission controls being just another skirmish. The older firms recognized that an extended political battle over NOx standards would be an irrelevant and futile distraction from the business of beating the competition on issues which customers cared about, such as ride comfort, engine noise and reliability. In the United States, the oligopolistic auto industry, which is renowned for its lack of innovation throughout that period, lashed out at a proposal which threatened the status quo and would have required substantial, and unfamiliar, design and development activity.[4]

---

[4] For a discussion of the American manufacturers' difficulties with innovation during this period, see James P. Womack, Daniel T. Jones and Daniel Roos, *The Machine That Changed The World*, Rawson, New York, 1990.

9.4.5. Commercial (Ir)Relevance of the Emission Standards

With hindsight, it seems that the requirements for three-way catalytic converters, and accompanying engine management technology, in one country and not in another had no significant impact on Japanese imports to the United States or US imports to Japan. Were the US manufacturers badly mistaken when they warned of the excessive costs of the new generation of catalysts?

Figure 9.1 shows the annual exports of passenger cars between the two countries, between 1970 and 1992. The first point to note is the huge ratio of Japanese to American exports. In any given year, Japanese exports to the United States range from 40 to nearly 1,000 times the export volume in the other direction. The reason for this is simple: the big US manufacturers have never taken any interest in Japan as a high-volume market. Until very recently, none of the big three bothered to offer right-hand drive versions of their cars in Japan. Nor did they try to cater for the average Japanese consumer's preference for smaller cars. Throughout most of this period, those US cars which were exported to Japan were mostly specialist vehicles such as sports models.

Japan's penetration of the US market, by contrast, was consciously planned and relentless, from the mid-1960s right through to the mid-1980s, when Japanese transplant factories in the United States began to meet an increasing proportion of the local demand for Japanese models. In the early 1970s, the US manufacturers were already shaken at the extent of the Japanese penetration of their market. In their minds they had failed to crack the Japanese market not because they expected Japanese drivers to sit on the wrong side of the car; not because two American cars would be unable to pass each other on a narrow road between rice paddies; and not because their take-it-or-leave-it sales pitch was insulting to consumers who were used to being nurtured and consulted by auto companies, but because of (mostly imaginary) 'non-tariff barriers'.

The latest and most dangerous 'non-tariff' barrier was, of course, the new set of vehicle emission standards. That is why the big American auto firms stirred up such a diplomatic fuss that they were given the right to sell dirtier vehicles in Japan between 1978 and 1981, than the Japanese themselves. Of course, this made no difference to their chances of selling more expensive,

Figure 9.1 Exports of Passenger Cars (x100)

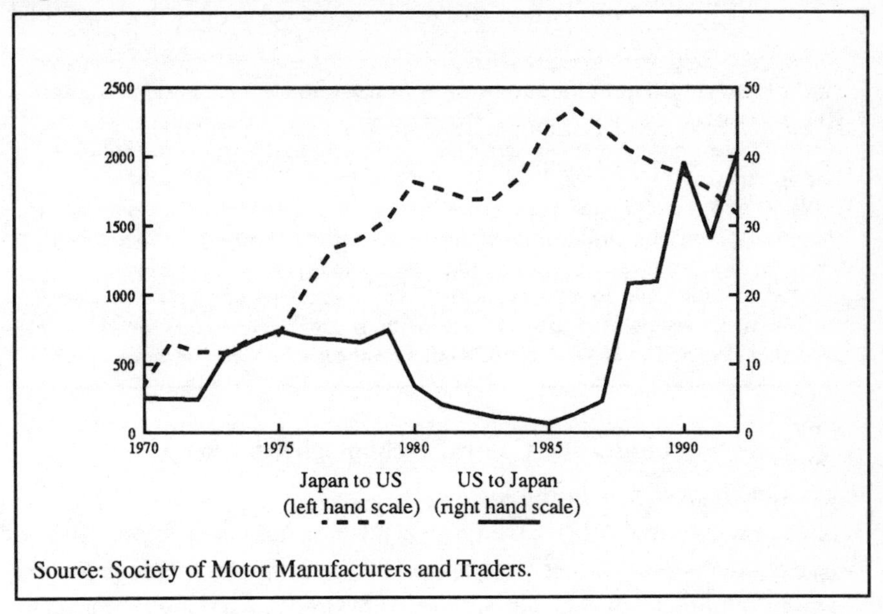

Source: Society of Motor Manufacturers and Traders.

poorer quality, inappropriate cars, but it pleased the US protectionists for a short while, although they would have preferred direct tariffs on Japanese exports to the United States.

9.4.6. Lean Production and Commercial Success

In the 1970s most Japanese car manufacturers had an underlying commercial advantage over their international competitors which enabled them to adapt more quickly to the new catalyst technology. Led by Toyota, they had developed the lean production system of manufacturing, which has by now effectively destroyed the mass production system in vehicle manufacturing and is spreading rapidly to many other volume manufacturing activities.[5] (See box overleaf.)

With the lean production system, Japanese firms were able to develop new models more frequently, make design changes more quickly and adapt to production changes within the factory with less disruption than other manufacturers. They took the introduction of three-way catalysts in their stride.

[5] A full description of the lean production system can be found in ibid.

Lean Production

At the assembly plant, lean production works by removing the 'safety net' provided by stockpiles of components in mass manufacturing plants. Assembly line workers, organized into cells, trigger new deliveries as they use up each new batch of components. Workers stop the assembly line when they find any problems. This is so inconvenient and costly that all problems, including component defects, are traced to their roots and eliminated, leading ultimately to 'right first time' and 'zero defect' manufacturing.

Design is also revolutionized. New vehicles are developed by dedicated teams led by respected, powerful individuals. Specifications are settled at an early stage, allowing major suppliers to take over much of the detailed design work. Body shapes are not changed after this point so work can begin on the slow, complex process of producing the dies which will be used to stamp out the body panels for the new vehicle once production begins. This can reduce the development cycle by a year or more.

9.5. Emission Limits and Control Technologies in Europe

9.5.1. Vehicle Pollution in Europe

With the exception of Athens, the European Union has not been prone to the severe photochemical smogs experienced by, for example, Los Angeles and Tokyo.[6] Pollution from vehicles was therefore slower in having an impact in Europe than in Japan or the United States. It took nearly 15 years for the EU to follow the lead of the other big two trading blocs in requiring catalytic converters. The political process leading to this outcome was uniquely European, and heavily influenced by concerns over competitiveness, both between European 'partners' and in relation to Japan and the United States.

Table 9.1 shows the progression of tailpipe emission standards in the EU. The first standards adopted by the European Community, in 1970,[7] were based on a regulation developed by a working group of the United Nations Economic Commission for Europe (UNECE), the ECE-15 Regulation of 1968. UNECE regulations were non-binding and based on unanimity among member states. By translating them into EC Directives, the European Commission was ensuring that EC member states could adopt the ECE standards as national law, without being accused of creating any trade barriers. These early EC directives were also optional for EC member states. This

---

[6] The climate in Athens, and consequently the air quality problems experienced there, most closely mimic those in Los Angeles or Tokyo.

[7] Directive 70/220/EEC, *Official Journal* of the European Communities L76, 6 April 1970.

Table 9.1: Gasoline Passenger Car Emission Regulation in the European Union

Directive	Year[1]	CO	HC	NOx	HC+NOx	Category
EEC/70/220	see note 2	25–55	2–3.2	–	–	Various,
EEC/77/102	see note 2	20–44	1.7–2.7	2.5–4	–	by
EEC/83/351	see note 2	17–33	–	–	6–8.7	weight
	1989(88)	6	–	1.6	0.8	>2,000cc
EEC/88/76	1993(91)	8	–	–	2	1,400–2,000cc
	1991(90)	11	–	1.5	4	<1,400cc
EEC/89/458	1992	4.6	–	–	1.3	<1,400cc
EEC/91/441	1992	2.72	–	–	0.97	All

All values are expressed in g/km.
1. Year in which the Directive becomes applicable to all new cars. Date in brackets is the year in which new *models* must comply.
2. Following the ECE regulations on which they were based, early EC Directives were non-binding. Implementation dates varied between member states.
Sources: N. Haigh, *Manual of Environmental Policy*, Release 5; OECD, *Transport and the Environment*, 1988

meant that there existed within Europe a patchwork of national emissions standards, as member states translated the various Directives into law at varying speeds, or even opted not to do so at all. For example, the United Kingdom did not incorporate the 1970 Directive into national law until 1976, when it did so through an amendment to the national Construction and Use Regulations.[8] This remained the pattern up to and including the 1983 EC Directive.

When the ECE regulation on which the 1983 Directive was based was being negotiated, a fundamental change of attitude towards auto emissions was taking place in (West) Germany. This was prompted by the rising concern over damage to forests as a result of acid deposition, especially SO_2 and NOx. Scandinavian complaints since the late 1960s of acidification of lakes had fallen on deaf ears, even in Germany, but had at least prompted some studies of the link between 'acid rain' and ecological damage. One project,

[8] J.M. Dunne, 'European Emission Standards to the Year 2000', Paper presented to the Institution of Mechanical Engineers Seminar on 'Worldwide Emissions Standards and How to Meet Them', 25–26 May 1993.

carried out by the University of Göttingen, had examined the effects of airborne pollutants on forest ecosystems in the Solling mountain range, in Lower Saxony. At the urging of the Federal Environment Agency (UBA), the result of this study – supporting a link between acid deposition and damage to trees – was published at the end of the 1970s, making acid deposition a subject of serious concern among scientists and policy-makers.[9]

Although the main culprit in acid deposition was thought to be SO_2 emissions from large industrial plants and power stations, government officials soon turned their attention to sources of NOx emissions, and noticed an alarming trend. Whereas SO_2 emitters were mostly stationary sources, and therefore relatively easy to control, road transport accounted for a substantial and rapidly increasing share of national NOx emissions. In 1970, private and commercial road vehicles accounted for 33% of NOx emissions, but by 1980 this had risen to 46%, or around 800,000 tonnes.[10] Traffic growth forecasts implied that road vehicles would soon be the major component of NOx emissions and might even undermine efforts to reduce NOx and SO_2 emissions elsewhere.

Prior to the creation of the German Environment Agency in 1986, the Interior Ministry (BMI) was responsible for air quality policy. Officials at the BMI were aware that one effective solution would be to draw on the forthcoming US standards, which could be achieved through the use of three-way catalysts (as Japan had been demonstrating for some time). However, they were sceptical of the prospects of achieving the switch from leaded to unleaded gasoline, which is essential for vehicles with catalysts. The main obstacle was the difficulty of persuading neighbouring countries, particularly other EC members, to accept unleaded fuel. If this could not be achieved, and Germany unilaterally made the switch to unleaded fuel and catalysts, German drivers would be trapped within their borders.

Policy-makers effectively dismissed the three-way catalyst option and considered alternative technological fixes, particularly the lean-burn engine. However, in 1982, the German auto manufacturers persuaded BMI officials

---

[9] 'Germany Case Study', in *Science Responds to Environmental Threats: Vol. II, Country Studies*, OECD, Paris, 1992.

[10] 'Umweltpolitik' (Environmental policy), German Minister for the Environment, 12/4006, December 1992, p. 34.

to execute an about-face on the issue. Germany exported cars to the United States in large numbers, and manufacturers such as Volkswagen were tooling up to begin production of export models fitted with three-way catalytic converters and the complex fuel injection systems required for efficient operation of the catalysts. The last thing they wanted was a separate technological approach in Germany and, ultimately, in Europe as a whole. They were unhappy about the prospect of tighter European standards, but if the German government was intent on this course, the best option would be for them somehow to persuade the rest of Europe to follow the US approach. As a potential bonus, German manufacturers might even have a head start and, they reasoned, a competitive advantage over European rivals.

It was this strategic industrial argument (combined with the need to do something quickly since the German public was in full cry over acid rain by 1983) which persuaded German officials to pursue the catalyst rather than the lean-burn approach in Europe.

9.5.2. German Tactics

The German strategy was assisted by a piece of great good luck which later proved to be largely responsible for neutralizing its biggest threat: the United Kingdom. There, airborne lead had become the big public environmental worry of the day. In 1980, a UK government working party recommended that the mean annual concentration of lead in the air should not exceed 2 micrograms per cubic metre.[11] The UK government was instrumental in reviving a stalled EC proposal based on this standard. Directive 82/884/EEC was adopted in December 1992.

In parallel with this, government attention focused on the lead content of gasoline. Existing EC regulations, dating from 1978, set an upper limit of 0.4 g/l and a *minimum* limit of 0.15 g/l. In April 1983, the Royal Commission on Environmental Pollution recommended that the government should seek the removal of the lower limit, and press for all new cars to be capable of running on lead-free gasoline. The United Kingdom, with German support, asked the Commission to draft suitable amendments to the 1978 Directive

---

[11] P. J. Lawther, *Lead and Health: the Report of a DHSS Working Party on Lead in the Environment*, HMSO, London, 1980.

and in June 1983 the European Council, under the chairmanship of Germany, decided in principle to reduce or eliminate lead additives. The end result was a 1985 Directive requiring member states to make unleaded gasoline available from October 1989, at the latest, and to reduce the upper limit on lead to 0.15g/l, at their own discretion.[12] UK public opinion was thereby placated and the German vehicle strategy was on course.

With the prospect of catalyst-equipped vehicles being free to travel throughout the EC from 1989, at the very latest, German tactics switched to manipulating vehicle emission standards in the EC and persuading consumers at home to buy vehicles equipped with catalysts. This second task was easily accomplished by providing tax advantages for vehicles equipped with catalytic converters, beginning in 1984. Public opinion was mobilized by strong environmentally tinged marketing on the part of the German auto manufacturers. Within a short time sales of catalyst-equipped vehicles took off, reaching one-third of new car sales in 1987, and passing the 50% mark the following year. The tax relief was withdrawn in 1991, when 97.1 per cent of new cars were equipped with catalysts.[13]

9.5.3. Regulatory Outcome

Despite the successful tax scheme within Germany, establishing strict European emission standards for NOx was clearly going to be the key to success for the German vehicle manufacturers. The traditional EC approach to vehicle emissions, where limits were not mandatory, no longer fitted the twin objectives of substantial NOx reductions and an EC-wide commitment to catalytic converters. A long battle with other European countries, and especially with the United Kingdom, Germany's ally in the effort to reduce lead, was about to begin.

Germany's first move, in 1983, was to threaten unilaterally to require cars sold in Germany to be fitted with three-way catalysts. Desperate to avoid this, the European Commission proposed a new vehicle emissions Directive in June 1984. The proposed emissions limits would require three-way catalysts. This was fiercely resisted by the United Kingdom, France and Italy. They argued that the extra costs of catalysts and engine management

[12] Directive 85/210/EEC, *Official Journal* of the European Communities L96, 3 April 1985.
[13] German Environment Ministry, Internal Document, 1993.

systems would disadvantage their smaller, cheaper cars relative to larger cars, on which German manufacturers tended to concentrate. They insisted that lean-burn engines were a better choice for Europe. Three-way catalysts also appeared to increase fuel consumption, by up to 16%, although switching from carburettors to fuel injection (essential for efficient catalyst operation) could cancel this out and even produce fuel savings, of up to 5%. However, lean-burn engines unequivocally reduced fuel consumption, by 15 to 22%.[14] Detractors of the three-way catalysts argued that their higher fuel consumption, combined with the expected shift in consumers' tastes towards larger, less efficient cars as the price differential between large and small cars was eroded, could outweigh the apparent emission advantages of three-way catalysts.

German officials, having recently been persuaded by their car manufacturers, on competitiveness grounds, to press for catalysts, confronted the objectors head on. In 1985, the Council of Environment Ministers reached a compromise which would have partly met the fears of Germany's opponents. Known as the 'Luxembourg Agreement', this set emission standards which would vary with size of car, so that only the largest class of car, with engine capacity greater than 2,000cc, would require a three-way catalyst.[15] Standards for smaller cars would be more stringent than previously, but were thought to be achievable with lean-burn engines. However, two of the smaller states blocked the Directive: Denmark because it wanted the strictest standards to apply to all cars, and Greece because it sought additional funds for environmental protection in Athens.

The impasse remained until a quite separate political initiative of the United Kingdom, the Single European Act, was ratified in 1987. This allowed a 'qualified majority voting' (QMV) procedure to be used for any measure which was required to implement a Single European Market. Under the new act, product standards were to be 'harmonized' throughout the EC. The European Commission considered vehicle emission limits to be product standards, and therefore QMV applied. Under the complex QMV rules a

---

[14] Commission of the European Communities, 'Report of the ad-hoc Group ERGA – Air Pollution', III/602.83 EN-Final, EC Commission, Brussels, 1993, cited in L.H. Watkins, *Air Pollution from Road Vehicles*, HMSO, London, 1991.

[15] N. Haigh, *Manual of Environmental Policy: the EC and Britain*, Release 5, 1994, 6.8–5.

minimum of one large country plus two smaller countries is required to form a blocking minority in the European Council. Denmark and Greece could no longer block the Directive, which was adopted in December 1987.[16]

Member states had agreed to look again at the smallest class of cars established in the December 1987 Directive, those below 1,400cc. In 1988, they proposed stricter standards for small cars, although these were also expected to be achievable without resort to three-way catalysts. The European Parliament had other ideas and used increased powers, which it acquired via the Single European Act, to tighten the standards, so that three-way catalysts would now be essential. The options for opponents of this move were now severely limited. The Council of Ministers could only amend the Directive to its original meaning by a unanimous decision, but clearly Germany, Denmark and others were delighted with the turn of events. The Directive would be abandoned altogether if a blocking minority could be formed, but none of the other countries were willing to side with the United Kingdom in such a drastic measure. The ultimate option would be for the United Kingdom to invoke the 'Luxembourg Compromise', but, sensibly, UK ministers took the only practical option and backed down.[17]

9.5.4. Commercial Outcome

So: doom, gloom and mutual recriminations in Britain; beer, bratwurst and soaring car manufacturer share prices in Germany. The long game was played out. Three-way catalysts would be required on all new cars from January 1993 and the ailing UK car industry would be dealt another blow, which it was surely ill-equipped to survive. But, unrecognized by the policy-makers, the United Kingdom's remaining independent volume car manufacturer had been quietly undergoing a remarkable internal transformation, which would demonstrate yet again the irrelevance of a crude linkage between emission standards and competitiveness.

---

[16] Directive 88/76/EEC, *Official Journal* of the European Communities L36, 9 February 1988.

[17] The Luxembourg Compromise (not to be confused with the Luxembourg Agreement) is the 'nuclear bomb' of European Community diplomacy. It was established by precedent by France, under General de Gaulle, and allows any EC decision to be held indefinitely in limbo if a member state feels that its vital national interest is at stake. It is only justified when the existence of the EC in its current form is under threat.

The UK car industry had been devastated by the postwar nationalizations which created British Leyland (BL), and the resulting management and union failures. However, many formerly independent brands of motor car retained some autonomy within the BL corporate structure. When the new Conservative government began to pursue its policy of divestment of publicly owned industries in 1979, it first had to rationalize the hopelessly loss-making BL. The Rover brand – BL's flagship luxury marque – was thought to have healthier prospects than most others. Over the course of several years, BL was reduced to a core of Rover cars and all-purpose vehicles (Land Rover and Range Rover) and the Metro small car, under the pre-nationalization 'Austin' marque. At the end of this process, BL was renamed Rover, and prepared for sale.

Another main plank of the Conservative government's industrial strategy was to encourage inward investment in the United Kingdom, particularly by Japanese firms eager to gain entry to the European market. In 1979, Japan's Honda car company began a programme of technical collaboration with the Rover business of BL. This cooperation expanded until finally Honda owned a 20% share of Rover, which had been purchased from the government by British Aerospace. New model development was shared between the two companies and Honda was transferring its world-beating lean production manufacturing and sourcing techniques to Rover.

When the time came to prepare for the introduction of catalytic converters, beginning with the 1993 model year, Rover was transformed. Where once it had been a struggling state-owned behemoth, now it was in the vanguard of lean manufacturing techniques within Europe. Consequently, tooling-up and component sourcing for catalytic converters was just another management challenge, which Rover was confident of achieving at lower cost than its European competitors, which remained stuck in the orthodoxies of mass production.

As had been the case a decade earlier with US–Japanese trade, the switch to three-way catalytic converters had no discernible effect on the relative success of Rover and the German firms. Other, more fundamental, forces were at work. German labour costs had risen alarmingly and labour productivity in the German car industry failed to keep abreast of the best in the world. Rover's relative price improved in the German market and word of the dramatic improvements in quality, thanks to the Japanese-inspired production techniques, got around.

Figure 9.2 compares Rover's exports of passenger cars to Germany with the total exports by German manufacturers to the United Kingdom. Rover's situation was similar to the US manufacturers' position in the Japanese market. Rover had a low-key distribution network which concentrated on more expensive models (including the Land Rover and Range Rover – specialist vehicles which are not included in these figures). Although this tends to make their sales highly volatile, it is clear that there was a strong improvement from 1989 to 1991, as Rover's improved quality and price competitiveness began to pay dividends.

Rover's success demonstrates once again the overwhelming importance of manufacturing excellence and managerial competence and, in particular, the power of the system of lean production pioneered by the Japanese. By greatly reducing the time and cost involved in developing new models, or making production changes in existing models, technical changes to meet new environmental regulations, or any other standards, have much less impact on the business. This is also true for any other volume manufacturing operation. As lean production spreads to other areas of manufacturing, policy-makers will need to understand industry's improved capacity to absorb and manage innovation, in order to set realistic goals for environmental, and other, improvements.

9.6. California's Electric Vehicles

Michael Gage, CEO of CALSTART,[18] has a vision: of a paradigm change in global vehicle manufacturing. He foresees that what is now a mature industry, dominated by a small number of giant manufacturers, will be superseded by a dynamic 'garage' industry, reminiscent of the early days of personal computers. He predicts that the catalyst for this change will be California's Zero Emission Vehicle regulations and the agent of change will be the electric car.

The alternative view is rather different. Even the best of today's electric cars suffer from one or more serious disadvantage in performance relative

[18] CALSTART is a California-based non-profit consortium founded to promote electric vehicles. Its member firms include major vehicle manufacturers.

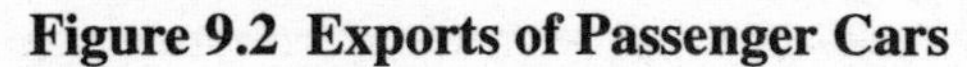

Figure 9.2 Exports of Passenger Cars

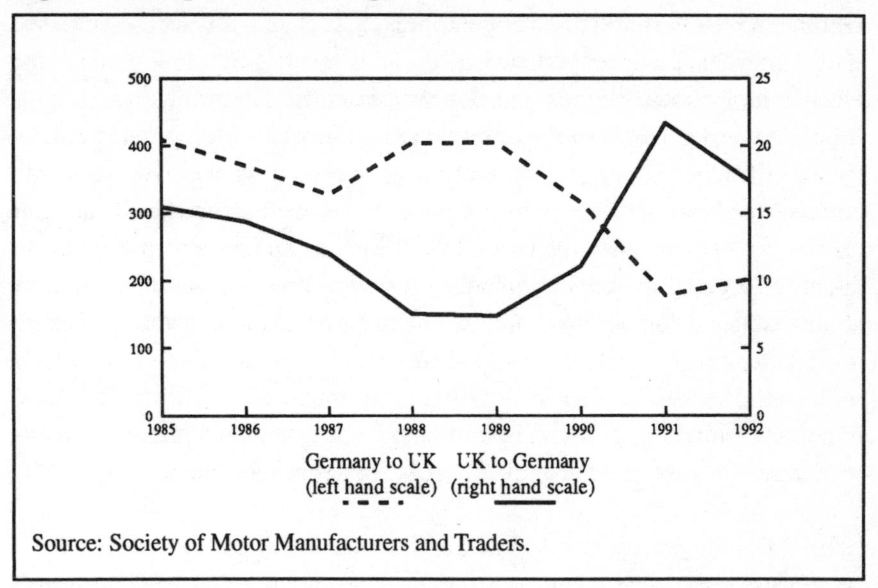

Source: Society of Motor Manufacturers and Traders.

to conventional vehicles. All of these problems, and the one indisputable advantage of electric vehicles, are nicely illustrated by Britain's much-loved and long-serving fleet of electric milk carts.

In the early hours of the morning, throughout Britain, bleary-eyed schoolboys and girls make their way to their local milk depot and climb aboard a small flatbed truck loaded with thousands of bottles and cartons of fresh milk. When their milkman starts the truck, something is missing – engine noise. These battery-powered milk carts make their way through the dawn streets without disturbing the sleeping residents, painlessly fulfilling the populace's deep-rooted need for fresh milk to be waiting on the doorstep when breakfast time comes.

Fortunately for the children who scurry from one house to the next delivering the milk, the milk carts travel very slowly: their maximum speed is around 30 miles per hour. Once back at the depot, the milk carts will not be needed again for another 18 hours or so, giving enough time for their batteries – seriously depleted after a journey of perhaps only 10 miles – to be recharged in readiness for the next day's deliveries.

Limited in range, slow, with long recharge times – but silent. These are the features which the public and popular press associate with electric vehicles, where they have any experience of them at all. In fact, new battery and electric motor technologies and lightweight construction techniques have improved performance to the point where reasonable driving range can be traded off against the kind of speed and acceleration we associate with conventional cars. Recharge times have been substantially reduced. But still no electric car can offer the overall performance and ease of use we have taken for granted in conventional cars for over 50 years. How has it come about that legislators in California are demanding that manufacturers produce these inferior goods in large quantities and, as seems likely, at greatly increased prices? The answer is a mixture of misleading information, false hopes and promises, political opportunism, and one man's grandiose plans to recapture a past 'golden age'. But first some background.

9.6.1. The Air Quality Problem in California

As discussed at some length in Chapter 8, the greater part of California's population lives in officially designated Air Quality Non-attainment Areas. That is to say, large parts of California, and in particular the Los Angeles basin, have unacceptably high levels of pollutants such as nitrogen oxides, volatile organic compounds and ozone. The major source of these pollutants is road vehicles. In the 1980s the political and social makeup of California excluded any possibility of restraining the growth in car use, let alone actually reducing it. The Los Angeles area air quality management authority had many incentive programmes to encourage commuters to share cars but it was clear that these would only have a marginal impact on the amount of car use.

The authorities' inability to control car use was vividly demonstrated when special lanes for 'ride-sharers' – vehicles with two or more passengers – were introduced. Initially, the existing outside, or 'overtaking', lane was reserved for ride-sharers, reducing the number of lanes available to single-occupant vehicles. This caused outrage among drivers who sat in even worse traffic jams, while one lane remained virtually unused. Rather than waiting long enough for drivers to modify their behaviour, and for ride-sharing to increase, the authorities caved in to public pressure. They 'added' an extra lane by converting former 'breakdown' lanes, or by squeezing all the other lanes slightly to create a new inside lane where inside breakdown lanes did not exist.

Anyone who has tried to use them will know how unpleasant these ride-share lanes can be. Ride-sharers are forced to drive in lanes which are often narrower than normal, are inches from the concrete wall of the central reservation, have rough, debris-strewn driving surfaces and are typically five or six congested lanes from the exits. For a while the extra lane may have eased congestion, but the long-term effect has almost certainly been increased pollution. By adding an extra lane the capacity of the highways has been increased, attracting more traffic and so increasing emissions of pollutants.

This weakness in the face of public demands for unrestricted vehicle use is combined with a conviction in California, and much of the rest of the United States, that restricting vehicle use harms the poorest sectors of society disproportionately. This has precluded apparently simple and effective remedies such as a uniformly enforced inspection and maintenance programme.[19] The politicians' concern for poorer drivers has also effectively ruled out any serious discussions of market mechanisms such as toll roads or congestion charging.

With these options written off, California has been forced to fall back on increasingly tight standards for tailpipe emissions for new vehicles, finally leading to the requirement for Zero Emission Vehicles (ZEVs).

9.6.2. The Zero Emission Vehicle Requirement

First, a word of explanation as to why a requirement for 'zero emissions' necessarily implies an electric vehicle, and why, in turn, emissions from electric cars are not really 'zero' at all.

---

[19] Annual inspections are required in California but are of limited effectiveness for at least two reasons. First, a vehicle failing the emissions test is exempted from any remedial action if such action would cost more than a certain sum. This allegedly prevents 'poor' drivers from being deprived of their cars, but creates an incentive for people with old, moderately polluting vehicles to allow them to deteriorate to such an extent that they avoid the need for any repairs. Second, the arrangement of having inspections carried out at general auto repairers has been criticized by the federal EPA and environmental groups on the grounds that these garages will be lax with their regular customers.

In theory, there are various ways of powering a vehicle without producing any harmful emissions. Fuel cells, for example, offer several options. They convert chemical energy directly into electrical energy without combustion taking place. A vehicle equipped with an appropriate fuel cell could be powered with natural gas or hydrogen. The only by-product from a hydrogen fuel cell is harmless water vapour. Unfortunately the fuel cell is still very much an experimental technology which has been put to practical use in only a tiny number of exotic, specialist applications, for example on spacecraft. Currently the only practical power source for a zero emission vehicle is an electric battery. Battery powered vehicles, as we already know from the story of the British milk carts, have been in use in appreciable numbers for some time.

The electric vehicle typically consists of some kind of battery and an electric motor, plus the conventional accoutrements of wheels, doors, chassis, etc. However it does not exist in a vacuum. It needs an infrastructure of recharging facilities, either based in dedicated sites analogous with gasoline stations, or at homes and workplaces. And the electricity for recharging has to come from somewhere. In general, this electricity will be supplied by the same mix of generating facilities which supplies domestic and commercial electricity consumers. In California this is a mix of hydro-electric, natural gas, geothermal, and (mostly out-of-state) coal fired power stations. Strictly speaking, therefore, the electric vehicle only produces zero emissions if the electricity is itself generated in a manner which does not produce any emissions. Unfortunately, there is no large industrial country where this is even remotely true. Indeed, in a country with a heavy reliance on coal fired power generation a wholesale switch to electric vehicles could cause a net increase in emissions such as carbon dioxide and sulphur dioxide. However, for California the best estimates show that replacing the *average* gasoline vehicle with an electric vehicle will cause a net reduction in emissions.[20] In any case, the pressing objective is to reduce emissions in urban areas, where there are too many cars but few power stations to worry about.

---

[20] For example, Tim Yau, *Gasoline vs. Electric Vehicles*, Electric Power Research Institute, 1990, predicts much reduced levels of VOCs, NOx and CO, but increased SO_2, from electric vehicles in California.

The Zero Emission Vehicle regulation which CARB promulgated in 1990 (and which was included in the 1990 Federal Clean Air Act as an option which other states can choose to adopt) was, therefore, clearly intended to relate to electric vehicles. It operates by setting a minimum requirement for the proportion of ZEVs which the large vehicle manufacturers must sell, by certain dates.[21] These requirements are:

- 2% of sales in 1998–2000;
- 5% of sales in 2001–2002;
- 10% of sales in 2003 and beyond.

The 2% target in 1998 is likely to be around 40,000 cars. This will apply to manufacturers selling 35,000 vehicles or more in the state each year. The manufacturers affected are likely to be Chrysler, Ford, GM, Honda, Nissan, Mazda and Toyota. From 2003, manufacturers selling over 3,000 vehicles will be included, thus encompassing most of the remaining European and Japanese firms.[22] Manufacturers who fail to meet their quota will face fines of $5,000 for each vehicle by which they fall short of the target.

Where will all these vehicles come from? Will they be only slightly souped up versions of milk carts? If they are to have prices and performance even approaching the price and performance of gasoline cars there will have to be major breakthroughs in battery technologies. Are these breakthroughs inevitable? If not, was this clear to regulators, back in 1990? As we shall see, the answers to these last two questions are 'No' and 'No'.

9.6.3. GM's Unfortunate Impact

Imagine, just for a moment, that I am a car salesman. I offer you a car which is faster, more comfortable and flashier than anything else on the market, at no extra cost. However, it does not exist yet and I want you to pay for it now.

---

[21] In belated recognition of the fact that the government cannot force purchases by consumers, CARB has reworded the ZEV requirement, to the effect that manufacturers must 'produce and deliver' ZEVs for sale. 'Rewording ZEV', *Electric Transport Information Centre Bulletin*, June 1993.

[22] California Air Resources Board Mobile Sources Division, *Technical Support Document: Zero-Emission Vehicle Update*, April 1994.

Would you buy it? Now imagine that you are responsible for vehicle regulations in California. A car salesman – in this case Roger Smith, the chairman of General Motors – offers you a car which will solve all your air pollution problems from vehicles. However, it does not exist yet (except in an experimental, commercially unacceptable form) and he wants you to give it regulatory and political support now. This is the implicit offer which the California Air Resources Board accepted in 1990. Perhaps this was forgivable, given the fact that the chairman of the world's largest car manufacturer was making concrete assurances that the car in question, the Impact, would be in large-scale production within a few years. But why, you might ask, was Roger Smith making such an extraordinary offer?

Superficially, the GM Impact started off as just another electric vehicle design project, no more likely to lead to a production vehicle than any of the previous and concurrent projects with which auto manufacturers throughout the world were involved. However, the then Chairman of GM, Roger Smith, had taken a close personal interest in this particular project since its inception. It had begun as a GM-sponsored solar-powered vehicle project, which resulted in an award-winning design, the Sunraycer. Roger Smith brought the Sunraycer designers to GM and teamed them up with engineers from GM subsidiary companies, Hughes and Delco-Remy.

Smith was enthusiastic enough about the Impact to insist on exhibiting the car at the Los Angeles Motor Show, early in 1990. The excitement and interest the public showed in the new car helped to persuade him to go ahead with a major development programme, a decision which was made public in April 1990. The California legislature ratified the California Air Resources Board's proposed ZEV regulation in September 1990. But what was Roger Smith's real motivation for committing himself and a substantial portion of GM's resources to a risky new technology which appeared very far from acceptability in the marketplace? It seems that the Impact was the cornerstone of a grandiose strategy through which GM would, with one mighty leap, transcend its dismal commercial performance of recent years and regain the world dominance it once enjoyed.

Along with the other big three American manufacturers, GM had been blasted by Japanese competitors throughout the 1980s. A Japanese car, the Honda Accord, even became the biggest seller in the United States. Faced with the very real prospect of terminal decline in GM's share of the car market, Roger Smith was prepared to contemplate a radical strategy involving a shift to an entirely new type of vehicle. This was a first-mover strategy on a truly epic scale.

At first sight, electric vehicles make a lot of sense for GM. Its electronics subsidiary, Hughes, provides necessary expertise in electronic engineering. Another subsidiary, Rockwell, has expertise in the lightweight materials required to allow electric vehicles to overcome their lack of range and power. Roger Smith probably imagined this happy coincidence of capabilities producing rapid solutions to the fundamental drawbacks of the electric vehicle. Indeed, the Impact prototype was hailed as having far better performance than any previous electric car, although still falling short of genuine usability.

With GM having a flying start in the key technologies, other vehicle manufacturers could be expected to struggle to catch up. With the mindset of an archetypal mass-producer, Smith might reasonably expect full-scale production of such a radical new technology to take at least 10 years to perfect. This represented the maximum cushion which he could establish between GM and its rivals, during which time GM could build an unassailable market share. After all, GM had come to dominate the postwar US car market because Alfred Sloan created a new kind of corporate structure and rivals such as Ford were slow to catch on. It was against this background, and with the knowledge that regulators at CARB were contemplating some kind of mandate for zero emission vehicles, that Smith made his commitment to the Impact in early 1990. From that moment, the decision by CARB to go ahead with a requirement for ZEVs was inevitable.

9.6.4. Putting the Shine Back into the Sunshine State

There was far more to the decision by CARB to propose a requirement for electric vehicles than a concern for the environment. Most dispassionate commentators agree that the ZEV regulation will do little to improve air quality – except in the very long term – and what improvements there are will be gained at enormous cost. (Some of the more effective and cheaper alternatives will be discussed later in this chapter.)

The ZEV regulation was heavily influenced by political and economic pressures. By 1990, California had slipped into a severe recession. One of the causes was the reduction in federal defence spending which began in earnest with George Bush's presidency. With a heavy concentration of defence-related manufacturers – among them GM's Hughes – the politicians were keen to encourage 'defence conversion'. This strategy was based on an acknowledgment that the heyday of US military spending had passed and that foreign customers could not be expected to replace them to any great extent. The only way to save thousands of high technology, highly paid jobs would be for the defence industries to find new, civil markets for their technologies and expertise.

The prospective electric car industry fitted the requirements for defence conversion perfectly. It was a high-technology business, with plenty of scope for continuing research and development. It was sufficiently large to provide a realistic alternative to the massive defence industry. And, since it could be dressed up as a necessary response to environmental problems, there were good prospects for massive federal aid for R&D.

The industry also satisfied another important need of state politicians in the United States. As highly autonomous political entities, with their own tax-raising and legislative powers, the states are in strong competition with one another and continually attempt to demonstrate their superiority as places to live and do business. California – the Sunshine State – had been riding high for most of this century, but had come down to earth with a bump at the end of the 1980s. What better way of reasserting its superiority than by stealing from Detroit the greatest industry of them all, the car industry?

The possible rewards of successfully creating an electric car industry in California are enormous. One estimate put the cost of building a single electric vehicle assembly plant at $500 million. Many such plants would be required to satisfy California's total demand of 1 million new cars every year.[23] CALSTART predicts that there will be 55,000 Californian jobs in electric vehicle components manufacture by the year 2000.

9.6.5. Lying for Fun and Profit

Clearly, then, the prospective electric car industry dovetailed neatly with the political agenda in California in 1992. However, this alone would not have

[23] A. Taylor, 'Why Electric Cars Make No Sense', *FORTUNE*, 26 July 1993, pp. 126–7.

been sufficient to persuade the regulators at CARB to propose the ZEV requirement to the state legislators. There was also a crucial failure in CARB's capacity to appraise the technical and commercial prospects for electric vehicles which caused it to over-react to GM's announcements on the Impact development programme. This failure has its roots in the distrustful, adversarial relationship between industry and regulator which is the norm in the United States, and which is easily discernible in the history of the relations between the vehicle industry and Californian regulators.

The vehicle manufacturers in the United States have exploited their superior knowledge of the technical feasibility of emission reduction measures in order to confuse and frustrate policy-makers' intentions. Policy-makers try to achieve parity in this information battle by recruiting from industry, buying into research and development efforts and talking directly to the manufacturers. The first option is perhaps the most effective, but the knowledge which an ex-engineer has of specific technologies will have a limited shelf life, and there are few opportunities for attracting such people from high earning industry jobs, except perhaps during recessions. Research and development is of only limited help as a source of information which might be relevant to regulation. Generally, the R&D sponsored by government agencies is concerned with speculative technologies which are still a long way from commercialization. Promising technologies move into the closed world of vehicle development activity, where the expenditures are far greater than in public research programmes. The third option – talking to the manufacturers – can do more harm than good. Manufacturers are likely to view an invitation from policy-makers to discuss the technical potential of emission reductions as an opportunity to mislead and confuse. As one official said with more resignation than bitterness, 'They sit across the table from us, smile, and lie to our faces'.

The advantage of the manufacturers over the regulators is demonstrated by the story of the development of the latest Californian vehicle regulations. The ZEV requirement is part of a larger body of legislation, commonly referred to as the LEV (Low Emission Vehicle) regulation. This regulation, which CARB adopted in 1990, sets a series of emission standards for vehicles. The standards for passenger cars and light vans are shown in Table 9.2.

Table 9.2: California Certification Standards for Light Vehicles (grams/mile)

Category	NMOG[1]	NOx	CO
1993	0.25[3]	0.4	3.4
TLEV[2]	0.125	0.4	3.4
LEV[2]	0.075	0.2	3.4
ULEV[2]	0.040	0.2	1.7

[1] Non-methane organic gases.
[2] TLEV = transitional low emission vehicle; LEV = low emission vehicle; ULEV = ultra-low emission vehicle.
[3] Adjusted for ozone reactivity.
Source: California Air Resources Board Mobile Sources Division, *Technical Support Document Zero-Emission Vehicle Update*, April 1994.

Manufacturers are free to produce vehicles conforming to these four types, or ZEVs, in any combination they wish, provided that their fleet-average emissions of a pollutant called 'NMOG' meet a certain standard. The NMOG target reduces over time and is shown in Table 9.3. In 1993, a manufacturer would have complied by producing vehicles which met the 1993 category alone. In 1994, at least some of the production had to meet TLEV, LEV, ULEV or ZEV standards. TLEV vehicle production alone will satisfy the standard in 1998, but in the following year something better will be required.

The curious feature of the regulation is this new regulatory target: NMOG.[24] It stands for 'non-methane organic gases' – basically just our old friend hydrocarbons, with the crucial difference that methane is deliberately excluded. A clue to the choice of this new category of pollutant as the driving force for all-round vehicle emissions lies in the full title of the regulation: 'Low Emission Vehicles *and Clean Fuels*'. The choice of NMOG was an attempt by CARB to force manufacturers and fuel suppliers to go down the route of 'cleaner' fuels, particularly methanol and compressed natural gas (CNG), or methane. Methane fuelled vehicles would have difficulty meeting a very tight target which encompassed methane. By deploying the argument that methane is less reactive and so less ozone-forming than other

[24] Other countries have targeted non-methane hydrocarbons, but this was the first time it had been adopted by American regulators.

Table 9.3: California NMOG Standard 1994–2003 (grams/mile)

Model Year	NMOG Fleet-Average Standard
1994	0.250
1995	0.231
1996	0.225
1997	0.202
1998	0.157
1999	0.113
2000	0.073
2001	0.070
2002	0.068
2003	0.062

Source: California Air Resources Board Mobile Sources Division, *Technical Support Document Zero-Emission Vehicle Update*, April 1994.

hydrocarbons, regulators were able to construct a new category of pollutant which could be used to discriminate between gasoline and 'clean fuels'. Having been assured by the auto companies that catalyst and improved combustion technology could not possibly meet an emission standard of 0.4 grams per mile for NOx and a substantial reduction in hydrocarbons, CARB officials devised the tapering NMOG standard in the full expectation that they were creating a future market for methanol and CNG.

To support the expected clean fuels market, agencies such as the South Coast Air Quality Management District (SQACMD) and the state Energy Commission supported demonstration programmes. Oil companies were encouraged to install methanol pumps in 50 gas stations. A 1991 regulation setting a NOx limit for commercial vehicles of 5 grams per hour, per brake horse power, was intended to force trucks and buses to switch from diesel to methanol. This was supported by subsidies for the Los Angeles Metropolitan Transit Authority to purchase a large alternative fuel bus fleet. Sadly, these efforts were confounded, as vehicle manufacturers first met the NOx standard with conventional diesel engines and then Ford told SQACMD officials that it expected to meet the ULEV standard for cars through improved catalyst and combustion performance in gasoline cars. If this is correct, the NMOG fleet-average requirement will never force the introduction of alternative fuels. Vehicle engineers later commented to SQACMD officials that they knew they could meet the standards, which they had formerly professed to be impossible, but were unwilling to make the effort unless forced to do so.

9.6.6. The Farce of Technology-Forcing

This word 'forced' has been a source of great misunderstanding and confusion throughout the history of Californian environmental legislation. Some officials are happy to be quoted publicly as saying that they use regulations 'to force technology'[25] and commentators frequently assert that this underpins the Californian approach. But what do they really mean by technology-forcing in this context?

For most people the phrase conjures up a vision of the authorities setting an aspirational target which cannot be met by existing technology (although there may be some research suggesting that it *might* be possible). The policy-maker, therefore, is setting out to 'create a vision'[26] to inspire/force industry to come up with new technological solutions under the dire threat that businesses will contract or even disappear if it does not do so. We might call this the popular version of technology-forcing. It is, of course, a myth.

As the above examples of Californian vehicle legislation demonstrate, what policy-makers aim for is *commercialization*-forcing. They try to establish what is technologically and economically possible and use that to guide their regulatory activity. In the case of the LEV and Clean Fuels legislation, they had established that methanol and CNG could achieve certain emission levels and believed (wrongly) that gasoline vehicles had no reasonable prospect of doing so. Their strategy was to force the commercialization of alternative fuels, whose technical performance in existing engine types was already well known. In this case they were misled by the manufacturers, with the result that the legislation seems certain to force the commercialization of improved gasoline vehicle technologies which the manufacturers were already reasonably confident of producing.

The ZEV case is different, and very unusual, in the sense that the regulators have been wrong-footed into adopting legislation which is genuinely technology-forcing, in the popular sense, although they were unaware of this at the time. Remember that CARB's decision to include the ZEV

[25] Bill Sessa, CARB spokesman, quoted in *Car and Driver*, 38, 8, February 1993, p. 12.
[26] CARB spokesperson, quoted in Taylor, see note 23.

requirement in the LEV regulation closely followed Roger Smith's announcement that GM would produce the Impact from 1995, in quantity and with the performance and at a price to persuade consumers to switch from gasoline cars. CARB regulators appear to have taken this at face value, assuming that Smith's market and technology assessment was correct. The ZEV requirement was, to their minds, a market-forcing instrument, a means of encouraging the commercialization of a newly available technology. Their relationships with the manufacturers were too weak and mistrustful to provide them with a counterpoint to Smith's highly personal view.

At the beginning of 1993, the first cracks started to appear in the CARB strategy. An upheaval at GM was the immediate cause. Roger Smith had retired and been succeeded as Chairman and CEO by Richard Stempel. In 1991 and 1992 GM ran up combined losses of over $12 billion, posting the biggest annual loss in American history in the process. The situation was so grave that in October 1992 Stempel was ousted by the GM board and replaced by Jack Smith, who immediately adopted an aggressive cost-cutting strategy. As part of this process the board ordered a reassessment of the Impact development programme. This resulted in GM dropping the full Impact programme in early 1993, in favour of the much more limited (and less expensive) 'PrEView' driver evaluation programme.[27] This precipitated negative press comment, as it became clear that a viable electric car was much less certain than CARB had originally believed.

9.6.7. Those Milk Carts Again

What are the real and prospective performance characteristics of the electric car? Consider the driving range first of all. GM claims a range of around 80 miles for the Impact. That is useless in greater Los Angeles. A typical journey there might involve driving from Hollywood to Diamond Bar, east of downtown, to visit the headquarters of the Air Quality Management District (SQACMD) – perhaps to check on progress in the fight against air pollution. Include one wrong turn in the downtown freeway system and the outward trip is over 50 miles. The return trip is about 40 miles, which leaves me walking for the last 10 miles, or waiting for 2–8 hours for the car to recharge,

[27] A plan to produce a small number of Impact-derived electric cars for consumers to test drive.

assuming that the recharging infrastructure is available. Anyone who tries to walk 10 miles in Los Angeles risks ending the journey in the local morgue. That is a fairly serious inconvenience to the average motorist.

Next, what about the refuelling, i.e. recharging, characteristics? There are three issues of interest: recharge costs, recharge time and supply (or infrastructure) implications. Let us examine the time and supply issues together, as the two are inextricably linked. To see the consequences of this new domestic appliance for the electricity supply industry, let us imagine a future where all Californian vehicles are electric (because, of course, they will be cheaper, cleaner, more fun to drive, etc., etc.). Suppose that in a worst-case scenario 5 million commuters return home within a one-hour period and plug their cars in for recharging. (There may be an important ball game on TV at 7pm.) CARB's cost estimates for electric vehicles assume a battery pack capacity of 27kW hours, so we will use that figure even though this is only sufficient for short-range vehicles such as today's Impact. Let us also assume that charging technology will have fulfilled some of the claims of fast recharge which we are already beginning to hear, and that this can be accomplished in 6 minutes. The charger is, therefore, rated at 270kW (this is several times greater than the total power rating of all the domestic appliances in a typical home). However, we can expect that our commuters are not all plugging in at the same instant. On average, only 10% of chargers will be running at any one time. Nonetheless, the total electricity demand is 135GW, or one-third of the total present generating capacity of the United States. No wonder the electric utilities are big fans of the electric car.

Of course, this expansion of electricity supply is simply unimaginable at present. Electric cars will need to be restricted to long recharge times if there is a risk of simultaneous recharging on the scale outlined above. This will put them at a permanent disadvantage to gasoline cars. But note that the advocates of the electric car, who claim that fast recharge times will be an important element in converting consumers from gasoline cars, contradict themselves when they claim that the EVs will be recharged gradually during the night, thus smoothing the demand for electricity and increasing the utilization of existing generating capacity.[28]

[28] CARB, *Technical Support Document*, p. 51, see note 22.

Running cost estimates for electric cars illustrate that advocates of electric vehicles frequently resort to wishful thinking of this kind. CARB contends that the fuel (electricity) costs of electric cars will be less than two cents per mile, compared with over four cents per mile for gasoline for the 'average' existing car. Here, CARB compares a highly aerodynamic experimental vehicle constructed using exotic lightweight materials, with a standard steel car where design and styling considerations outweigh aerodynamics. An equally exotic gasoline car would also achieve running costs of less than two cents per mile. Indeed, some small cars currently on the market already achieve this. Why do consumers not buy these cars in greater numbers? The answer is that, in the United States at least, consumers don't care about fuel costs. What they want is performance, something an EV comes nowhere near to giving them. They will be equally unimpressed by the fuel price 'advantage' of EVs.[29]

9.6.8. Jobs for the Boys

So what are the political prospects for the electric vehicle? In May 1994, the ZEV requirement passed a mandatory technical and economic review, despite GM's near-abandonment of the Impact and the lack of any fundamental technological breakthroughs in the two years since the last review. Part of the explanation is that the hopes of defence conversion and reasserting California's competitiveness relative to other states still carry a lot of political support. In addition, the hoped-for windfall of federal and private support for research and development has come about. Most notably, the US Advanced Battery Consortium (USABC), which is joint-funded by the big three manufacturers and the federal government, will spend $260 million on battery technology research, over four years.

The electric vehicle legislative path has been locked in by the hype and over-optimism which followed the 1990 decision. Other states appear to have narrowly escaped tying their hands in this way. After initial enthusiasm for the Californian ZEV requirement, it now seems likely that legislation elsewhere will merely allow ZEVs as one option for manufacturers to use in achieving fleet-average requirements for NMOG. For example, the Ozone

[29] CARB's figures are based on special reduced electricity tariffs for EV charging. These tariffs will not persist if EVs become the single biggest electrical appliance, in terms of consumption.

Transport Commission (OTC), a consortium of regulators from north-eastern states with a shared characteristic air mass (i.e. they pollute one another), is seeking approval from the federal EPA for a proposal which would allow the 2003 NMOG standard to be met through sales of 63% LEV type vehicles and 37% ULEV type vehicles, with no mandatory requirement for ZEVs.[30]

9.6.9. Toys For the Boys

Electric vehicle technologies have developed rapidly since 1990, at least within their fundamental limitations. In keeping with the vision of Michael Gage, CEO of CALSTART, much of this innovation has occurred within small, start-up 'garage' firms. But the direction of that innovation is moving away from his ultimate vision of small EV specialists exploiting low entry costs and simple manufacturing requirements to overthrow the auto giants.

To take just one example of the new 'garage' companies, consider Renaissance Cars and their new vehicle, the Tropica. This is a small, two-seater run-around with a range of 65 miles and top speed of 65mph. It is extremely light, thanks in part to the designers dispensing with a roof or air conditioning. For each Tropica sold after the ZEV mandate becomes operational, Renaissance Cars will earn a credit. Bob Beaumont, founder of the company, expects the big manufacturers to purchase these credits from him for a significant proportion of the $5,000 fine they would otherwise face for not supplying their own EVs.[31] With a subsidy of several thousand dollars each, Renaissance Cars may be able to reduce the price of its little run-around to the $5,000–$8,000 range.

What sort of market niche might such a vehicle fill? It would be perfect for parents to buy for their children of high school age: too slow to be very dangerous, yet acceptable to image-conscious kids because of its green credentials. Get the design and branding right and it could be a runaway success. Others have spotted similar opportunities. The El-Jet, from Danish manufacturer Kewet, is intended for niche uses only, such as local shopping trips. Added together, these niche markets might even exceed the 1998 target of 2% of Californian sales.

---

[30] 'EPA will approve OTC petition but still seeks compromise', *Hart's Octane Week*, 9, 31 October 1994, p. 43.

[31] 'Electric Cars', *International Business Week*, 30 May 1994, pp. 36–42.

Which apparently happy thought brings us to the great, gaping hole in the Californian strategy – what is a market, and, more fundamentally, what is a car? Take the example of the Tropica. This is a vehicle type for which no precedent exists and which may well create for itself a market niche which did not previously exist. This may legitimately count towards filling the ZEV quota, but where is the benefit for air quality? Of course, there is none. The ZEV regulation assumed, simple-mindedly, that an electric car would be an exact substitute for a gasoline car. We can see already that this is unlikely to happen for a very long time, if at all. Instead, what has been created is an incentive to sell 2% more vehicles in 1998 than would otherwise have been the case. That $5,000 fine, multiplied by an estimated 40,000 units, gives GM, Ford, Chrysler and the rest a $200 million incentive (rising to $1 billion per year in 2003) to persuade consumers to buy new electric toys for their kids' amusement or perhaps for occasional shopping trips. For the big manufacturers the alternative is to subsidize expensive electric versions of existing models sufficiently to persuade consumers to forgive their lack of functionality, thus cannibalizing sales and profits of their gasoline models. Which option would you choose, faced with the over-arching necessity of satisfying your shareholders?

Ultimately, California's air quality problems stem from a political inability to restrain demand for motor transport. That basic impotence has been enshrined in the ZEV requirement, reducing its effectiveness as a means of cleaning the air.

9.7. Conclusions

The car industry is taken extremely seriously by most governments. Protective attitudes in bureaucracies and among politicians raise the possibility of special treatment for the industry when social policies such as safety and the environment appear to threaten its interests. The history of vehicle emission regulations since the 1970s reveals that efforts to influence policy with arguments about cost and competitiveness have had varying degrees of success.

One reason for these differences is the independence and political credibility of the environmental policy-maker. In Japan the metropolitan authorities took the lead in protecting their citizens' interests when the national bureaucracy, where social and industrial objectives are more likely to be conflated, had been content with the major manufacturers' initial incorrect portrayal of the technical potential for NOx reductions.

Information has provided another source of influence over emission regulations. Japanese officials and manufacturers openly discussed the technical potential of three-way catalysts. This dialogue was later institutionalized in annual technical hearings between the national Environment Agency and the manufacturers. In California, by contrast, there is strong political will behind environmental policy, but poor dialogue. The relationship between policy-makers and the manufacturers is adversarial, with the manufacturers exercising their power through superior information. One consequence of this is that the LEV regulation failed to create the market for alternative fuels which the Californian regulators had been led to expect. No doubt the manufacturers are meeting the new emission standards at lowest cost to themselves by refining catalytic converters for gasoline vehicles rather than developing alternative fuelled vehicles. However, the Californian policy-makers were aiming for a certain policy outcome when they adopted NMOG as their key emission standard. In doing so they probably compromised on the levels of other emissions. If they had known what the industry knew, they might have had a different set of priorities. They also put a lot of wasted effort and taxpayers' money into research and demonstration connected with alternative fuels.

Environmental policy-makers take great pains to ensure that the objectives they write into legislation and regulations are achievable. They desperately wish to avoid situations where their policies are dependent on scientific or technical breakthroughs which may or may not appear. However, they are very often in the business of forcing the commercialization of technologies. Often, they need a good trusting dialogue with industry in order to distinguish between policies which would force technology and those which would force commercialization of existing basic technologies. The 1978 Japanese standard for NOx was not technology-forcing. It was based on demonstrations of an existing technology. The standard was simply appropriate, in the sense that it would contribute to Japan's air quality objectives, and achievable, given the true state of the technology at the time.

The electric vehicle requirement in California shows what can happen when the relationship between the policy-maker and industry is so poor that policy-makers cannot discriminate between technology-forcing and commercialization-forcing. Typically, an uncertain regulator behaves conservatively. In California, hopes of jobs and improved competitiveness at a time of recession led policy-makers to make a less than cautious interpretation of GM's announcement of the 'Impact' electric car. If there are no major technical breakthroughs, electric vehicles will remain unacceptable as substitutes for today's passenger car. On a narrow interpretation the ZEV target might then be met through new niche markets, but the policy-makers' original intention – replacement of a proportion of the expected market for gasoline vehicles with electric vehicles – would not be realized.

Differences in the vehicle manufacturers' own abilities to cope with change, including changes required by emission regulations, have also influenced the course of policy. Lean production gave the Japanese manufacturers big advantages over the United States companies in design and new model development. Firms such as Honda, Toyota and Nissan had less reason to fear the production implications of new catalytic converters and engine management technologies in the late 1970s than their American counterparts, and so less reason to lobby policy-makers. The big US manufacturers have since adopted many of the major features of lean production.

Europe has been slow to adopt lean production, just as it was slow to adopt mass production in the vehicle industry. United Kingdom policy-makers were fighting for Britain's car industry, i.e. Rover, on the premise that the fundamentals of its manufacturing operation, and therefore much of its costs, were similar to other European firms. Their negotiating stance during the 1980s might have been very different had Rover already been a lean producer.

Policy-makers need to understand technological developments and, more fundamentally, the nature of the industry they are regulating. To achieve this, they need industry's cooperation. (The political and institutional means of ensuring cooperation are discussed in more detail in the country studies.) Well informed, competent policy-makers are in a better position to achieve their objectives through mechanisms which make use and take account of firms' capacity for innovation.

Chapter 10

Cleaning Coal

10.1. Introduction

Coal has powered the Industrial Revolution. It remains the dominant fuel for electricity production and is expected to outlast oil and gas supplies by hundreds of years. Not coincidentally, coal burning has been a root cause of some of the major pollution issues in the industrialized world. When agricultural reform increased the supply of labour, and coal-powered industries attracted that labour to new cities, urban air pollution was created. Attempts to control that pollution were the impetus for much of the world's environmental legislation, up to the 1970s.

During the 1970s and 1980s acid rain became the defining issue of environmental politics. A whole new set of concepts entered the public consciousness as a result. Pollution was now transnational. It was being imported and exported. It was destroying vast tracts of pristine wilderness. It was intercontinental, even global.

But acid rain was never going to be eliminated at the stroke of a pen. Coal's dominant position within energy generation, the importance of energy within the economy and the 'strategic' status of power engineering firms in the largest economies ensured that acid rain was as politically charged for economic and industrial reasons, as it was for environmental reasons.

This case study traces the interwoven strands of environmental legislation, pollution control technologies and national policies towards technology suppliers. It begins with the major national efforts to control urban air quality, and tracks the consequences of these efforts for later, international, efforts to control acid rain. For simplicity, it concentrates on the control of emissions of the major acid rain gases, sulphur dioxide (SO_2) and nitrogen oxides (NOx), from sizeable boilers, notably those used in power generation.

10.2. The Problem with Burning Coal

Take a host of living organisms, rich in hydrocarbons and in the exotic trace elements required to sustain their lives. Crush them under layers of rock, over a long enough time for geological processes to interleave more metals and compounds with the original biological material. Then blast and hack it out of the ground, pile it up and set fire to it. The resulting cloud will include particles of rock carrying metals and unburnt hydrocarbons, metal vapours such as mercury, and combustion products such as SO_2, NOx, dioxins and carbon monoxide and carbon dioxide. The role of technology in coal power generation has been to extract the maximum amount of useful energy from the coal, while minimizing the amount of pollution being produced and released. This is a complex task.

10.2.1. Effects of SO_2 and NOx

Local Effects In the past, the worst urban air quality problem was the combined effect on human health of sulphur dioxide and suspended particulates. In cities, the burning of coal in power stations, on industrial sites and in homes, produced sulphur dioxide and particulates, and other industrial processes contributed more particulates. Adverse effects on health were worst in cold, stable weather conditions. The suspended particulates provided nuclei on which water vapour condensed, forming – now polluted – fog. The sulphur dioxide gas dissolved in the water droplets, producing sulphuric acid. The resulting acidic, polluted mist is smog.

During smog episodes, children, the elderly and those with existing respiratory and heart disease are particularly vulnerable. The most severe smogs are directly linked to hundreds and sometimes thousands of excess deaths among these groups of people.[1]

Even at the very much lower concentrations of SO_2 and particulates generally found today in western cities, vulnerable groups such as asthma sufferers are adversely affected. The World Health Organization estimates that the condition of those with respiratory disease will worsen if they are exposed to SO_2 or particulate concentrations of 250 $\mu g/m^3$ or more over a 24-hour period.

[1] In one severe London smog particulate levels up to 6,000 micrograms per cubic metre ($\mu g/m^3$) and SO_2 levels up to 4,000 $\mu g/m^3$ were recorded.

At high concentrations, SO_2 also causes severe erosion of common building materials, such as limestone, sandstone, slate, mortar and metals, and causes bleaching and other damage to plants, reducing their growth rates.

Regional and Global Effects: 'Acid Rain' The presence of SO_2 and NOx in the atmosphere makes rain more acidic, sometimes up to 30–40 times the unpolluted acidity.[2] Although acid rain is the commonly used term familiar to the public, dry deposition of SO_2, NOx and their compounds – sulphates and nitrates – exceeds the rate of wet deposition through precipitation, in many areas. We will, however, stick with the label 'acid rain', denoting wet and dry acid deposition, for simplicity.

In western Europe, a relatively small area including parts of Belgium, the Netherlands and Denmark was experiencing severe acidification by 1959. As coal power generation in Europe accelerated, the affected area expanded to include northern France, and much of West Germany, Sweden, Norway and eastern England, by 1966. Swept along by prevailing weather patterns, acid rain is truly transboundary. Emissions from some US power plants precipitate partly over Canada, before crossing the Atlantic and adding significantly to acid rain in Europe.

Acid rain's effects on ecosystems are extremely complex. Early, simplistic conjectures have been rejected or modified time and again. The search for a firm scientific understanding of its effects, and the difficulty of separating those effects from unrelated factors which harm ecosystems, is a major part of the acid rain story. These scientific aspects and their political implications have been written about elsewhere, and are not considered here in any great detail.[3]

Acid rain is now understood to have the greatest effect in areas where it overcomes the natural capacity of soils and lakes to neutralize, or 'buffer', acidity. This capacity depends largely on the geochemistry of the underlying bedrock. Igneous and metamorphic bedrocks are low in the calcium and magnesium which buffer acidity and are therefore susceptible to acidification.

---

[2] D. M. Elsom, *Atmospheric Pollution: A Global Problem*, 2nd edition, Blackwell, Oxford, 1992, pp. 83–4.

[3] See, especially, S. Boehmer-Christiansen and J. Skea, *Acid Politics*, Belhaven, London, 1991.

Increased acidity has direct effects on plant life, such as forests, or fish in lakes and rivers. In addition, metals such as aluminium become mobilized in acidic conditions, important chemicals leach out of soil, nutrients become less readily available to plants and animals, and the composition of species of essential micro-organisms changes. All these factors contribute to more visible effects, such as dying forests and fish populations.

10.2.2. Controlling SO_2 Emissions

Minimizing Sulphur Input The simplest way to reduce emissions of SO_2 from coal burning is to burn coal with a low sulphur content. In the United Kingdom, the sulphur content of hard coal varies from 0.7 to 2.0%. This is generally considered to be a 'high' value. But brown coal, or *lignite*, often has a very high sulphur content. Austria produces brown coal with a sulphur content as high as 7%.[4] Brown coals have a lower energy content than hard coal so more must be burned to produce the same output of energy. This further increases the typical SO_2 emissions associated with using brown coal.

Some countries began to address the problem of SO_2 emissions by setting maximum allowable sulphur contents for coal. However, widespread switching to low-sulphur coal has been limited by lack of availability and higher prices, by political protection of mining interests in high sulphur coal-producing areas and by the inability of even low sulphur coal to meet emission limits set out in regulations.

The sulphur content of coal can be reduced further by removing part of it, i.e. cleaning the coal. Simply by crushing and washing it in water, up to 50% of the sulphur can be removed from most coals.[5] Once emission standards make expensive post-combustion control technologies essential, coal cleaning and the initial sulphur content tend to become less important, and can be more expensive than slightly increasing the efficiency of the control technology.

Flue Gas Desulphurization Flue gas desulphurization (FGD) is the dominant technique for reducing SO_2 emissions from coal burning. The flue gases, consisting of all the waste gases from the combustion process, are

---

[4] J. L. Vernon, 'Market impacts of sulphur control', IEA Coal Research, IEACR/18, October 1989.
[5] Ibid.

reacted with a slurry, solution or spray, containing an alkaline sorbent. The product of the reaction between the sorbent and the SO_2 is removed before the remaining gases are allowed to escape to the atmosphere.

Most FGD units world-wide are of the 'wet scrubbing' type, using a slurry or solution. The most common sorbent is limestone, which reacts with the SO_2 to produce calcium sulphite, which must be disposed of, or calcium sulphate (i.e. gypsum), which can be sold as a building material.

Spray-dry scrubbers, most of which use slaked lime as a sorbent, are less common. The flue gases pass through the spray of sorbent, producing a mixture of calcium sulphite/sulphate and fly ash, which cannot be used commercially. Sorbent injection, where the sorbent is injected into the boiler, is rarely used and produces a similar mixture of reaction products.

Clean Coal Technologies For decades, nearly all large coal power plants have been based on the pulverized fuel firing (pf) design. In the pf plant coal is crushed to a fine powder and blasted into the boiler in a stream of air, through a series of burners. The FGD technologies discussed above apply to this design of plant.

Radically different combustion plant designs, collectively known as 'clean coal technologies', offer very different ways of controlling SO_2 emissions. In a fluidized bed combustion (FBC) plant, the coal is fed into an incandescent, turbulent mass of inert particles. Traditional sorbent, such as limestone, can also be fed into the bed, achieving 75–95% sulphur removal.[6] The inclusion of this sulphur removal capacity in an FBC plant involves very little additional capital expenditure.

In the integrated gasification and combined cycles (IGCC) plant, coal is partially combusted to produce gaseous fuel, which feeds a gas turbine. The gasification process is manipulated by strictly controlled introduction of oxygen, steam or hydrogen. Choosing just the right ratio of these gases will cause nearly all the sulphur to form hydrogen sulphide (H_2S) gas. It is a relatively simple matter to extract the H_2S and produce pure sulphur, which can be sold for industrial uses.

---

[6] Walter C. Patterson, *Coal-Use Technology: New Challenges, New Responses*, Financial Times Management Report, London, 1993.

These elegant methods of dealing with sulphur are beginning to make clean coal technologies commercially attractive. Until recently, high initial capital costs made them uncompetitive with new pf plants. Now, the additional capital cost of high efficiency FGD equipment and other pollution control measures required for pf plants is beginning to erode their advantage over the clean coal technologies.

10.2.3. Controlling NOx Emissions

Formation of Nitrogen Oxides During coal combustion NOx is formed via three distinct processes:

- Fuel NOx is derived from nitrogen-bearing compounds in coal, and accounts for around 75% of NOx formation in a typical pulverized fuel power plant. When oxygen is abundant, these compounds will react to form NOx. When oxygen is scarce, i.e. in fuel-rich combustion conditions, the compounds react to form nitrogen gas.
- Thermal NOx is the term given to the NOx formed when nitrogen gas reacts with oxygen. It accounts for around 20% of NOx formation and occurs at a significant rate at temperatures above 1,500 degrees Celsius. Thermal NOx formation can be minimized by limiting the combustion temperature, increasing the rate at which combustion gases flow through high temperature zones or restricting the available oxygen. Unfortunately, all these measures will tend to reduce the completeness of combustion and the overall efficiency of power generation.
- Prompt NOx, a poorly understood process involving highly reactive hydrocarbon molecules, accounts for only around 5% of the total.

Minimizing NOx Formation Measures to reduce NOx formation are all based on altering combustion conditions in ways which minimize, primarily, fuel NOx formation. The most important technology for achieving this in conventional pulverized-fuel boilers is the low-NOx burner. The burner is the nozzle projecting through the furnace wall which introduces the pulverized coal into the furnace. A simple burner fires a mixture of coal and air into the furnace. Combustion begins at the nozzle itself. A low-NOx burner is designed to control the delivery of air in such a way that combustion takes place in stages. In the initial stage the nitrogenous compounds are gasified and then react in a low-oxygen environment, forming nitrogen gas in preference to NOx. Secondary or tertiary air flows, supplied separately from the coal-bearing central flow, complete the combustion of the fuel.

The design of the boiler itself can affect fuel NOx formation. For example, tangential-fired boilers, with the burners located in the corners of the furnace, have lower NOx formation than wall-fired boilers. However, with control measures, the final levels of NOx formation are similar for these two designs. Other design modifications, such as air-staging in the furnace and fuel-staging, can substantially reduce emissions.

NOx Removal Technologies Removal of NOx from flue gases can be achieved in two main ways: selective catalytic reduction (SCR) or selective non-catalytic reduction (SNCR).

Selective catalytic reduction is similar in principle to the reduction of emissions by catalytic converters fitted to vehicles. The flue gases pass through layers of the catalyst material, where the reducing reaction takes place. However, whereas in vehicles the NOx reacts with other gases already present in the exhaust, SCR requires the addition of ammonia to the flue gases. The NOx reacts with the ammonia to produce nitrogen gas and water. The most common catalyst material is titanium oxide. This was developed in Japan and and is much more common than other catalyst types, such as zeolite or activated carbon.

Selective non-catalytic reduction involves spraying ammonia or urea into a specific temperature zone of a boiler, where significant NOx formation is taking place. The NOx reacts with the ammonia to form nitrogen and water, reducing NOx emissions by 30 to 70%. This technique depends on achieving good mixing between the combustion gases and the ammonia or urea. The upper limits on achievable NOx reductions have been raised in recent years due to improved modelling of combustion in furnaces, but factors such as the variation of combustion conditions in relation to the furnace output and the inherent random variability of the distribution of gases in the furnace continue to restrain the effectiveness of SNCR.

10.3. Local Impacts of Coal-burning: Historical Experience

10.3.1. Germany

Germany, a major industrial power for nearly two centuries, provides a long record of the adverse impacts of coal burning.

Forest damage, the most visible sign of acid rain in recent years, has a historical precursor in smoke damage to trees in and around towns and cities. In the nineteenth century, German concerns over this led to the creation of an international committee of smoke damage experts. As early as 1850, the committee identified high concentrations of SO_2 from industrial facilities as the major noxious substance involved in 'smoke' damage.[7]

However, this damage was officially tolerated as an unavoidable consequence of industrialization. In the Ruhr Valley region, attempts had been made to counter the problem by planting smoke-resistant trees and trees with a low sensitivity to smoke. Clean air policy was focused on preventing damage to human health, not protecting ecosystems. As in other countries, especially Britain and Japan, the principal concern in Germany was urban smog. The main pollutants contributing to smog were particulates, which were reduced by installing filters and precipitators, and SO_2, which was tackled by constructing high chimneys to disperse the gas into the atmosphere. (Particulates still settled locally, hence the need for dust removal technologies.)

10.3.2. The United Kingdom

Air pollution control began early in the United Kingdom. King Edward I (1272–1307) issued a proclamation banning the use of sea-coal (collected from the shoreline) in open furnaces. A first offence invited a fine; for a second offence the offender's furnace would be demolished; and a third was punishable by death.[8] A more lenient attitude over the next seven centuries led to the London 'smogs' of 1873, 1880, 1881, 1882, 1891, 1892, 1901 and 1942, culminating in the disastrous smog of 5–8 December 1952, in which 4,000 people were killed.

---

[7] *Sondergutachten Waldschaden/Luftverunreinigungen*, Federal Council of Environmental Advisors, 10/113, Bonn, May 1983, p. 8, cited in *Science Responds to Environmental Threats: Country Studies*, OECD, Paris, 1992, p. 186.

[8] W. Martin, 'Legislative Air Pollution Strategies in Various Countries', *Clean Air*, 9, 1975, pp. 22–8, cited in Elsom, p. 238, see note 2.

The UK urban smogs were largely the result of smoke and SO_2 emissions from domestic coal fires. The 1956 Clean Air Act concentrated on this source, giving local authorities power to establish smoke control areas, where only smokeless fuels could be used, and awarding grants to householders for conversion to electricity, oil, gas and anthracite. The act reinforced existing social changes, such as increased demand for central heating with rising living standards and the collectivization of local authority tenants into high-rise buildings, both of which reduced domestic coal use.

However, public attitudes to power stations were coloured by a legal battle in the 1920s between a Manchester electricity company and a farmer who claimed a nearby power station was damaging his crops. When the London Power Company (LPC) proposed to build Battersea Power Station in central London, public objections led to strict conditions being imposed on smoke and SO_2 emissions. The LPC was faced with developing an SO_2 removal process. Experiments and pilot plants demonstrated the feasibility of scrubbing the flue gas with alkaline solutions. In 1933, Battersea was commissioned with a flue gas desulphurization system using Thames river water with added chalk (later alkaline sludge).

The first Battersea FGD plant achieved 90% SO_2 removal but was expensive to maintain due to high corrosion.[9] Later plants at Battersea, and then Bankside Power Station, improved on the early designs but the Central Electricity Generating Board (CEGB) still considered FGD to be a troublesome and expensive technology.[10] Its work on the design of chimneys demonstrated that a plant's emissions could be dispersed at a high level, reducing ground level concentrations to an acceptable value. This, combined with a strategy of siting power stations outside cities so that strong public opposition was unlikely, led to the CEGB abandoning FGD technology.

These combined measures had the desired effect, bringing substantial reductions in urban smoke and SO_2 concentrations by the middle of the 1970s, by which time the issue of smoke and SO_2 meant very little to the public. When the problem next returned to the United Kingdom the solution would

[9] J. Bettelheim and W. S. Kyte, 'Fifty Years' Experience of Flue Gas Desulphurisation at Power Stations in the United Kingdom', *The Chemical Engineer*, June 1981.
[10] Boehmer-Christiansen and Skea, see note 3.

once again be based on a massive decline in coal use, in preference to technological fixes, this time in the power sector.

10.3.3. Japan

Japan's early phase of industrialization, which began with the Meiji Restoration in the mid-nineteenth century, brought with it pollution just as severe as any encountered in Europe. In the early 1900s, extreme SO_2 pollution from the smelting plants around the Ashio Copper Mines caused protests of ill-health from local residents, and killed all vegetation within the immediate vicinity. Occasionally such incidents prompted action by local officials but there was no equivalent of Germany's or Britain's national legislation on industrial processes.

After the Second World War, Japan concentrated on rebuilding its industries. The postwar period of rapid expansion led to severe industrial pollution problems during the late 1950s and the 1960s. Some of the more notorious pollution events from this period are documented in the introduction to Chapter 7. One of these was largely responsible for prompting a nationwide attack on industrial SO_2 emissions which went far beyond anything seen in other industrial countries. Unlike the situation in the United Kingdom during the smog period, Japan's urban SO_2 emissions were dominated by industrial sources which were often in very close proximity to residential areas. This effort was to lead eventually to the development of several of the technologies now most commonly used to deal with emissions from coal power stations.

The turning point in Japan was not a dramatic killer smog, as in the United Kingdom, but a health disaster which developed over several years. In 1959, the Yokkaichi petrochemical plant began operation. By 1963, nearby residential areas were regularly experiencing SO_2 concentrations of over 0.5 ppm. Before long, nearly 1,200 residents were identified as suffering from 'Yokkaichi asthma'. In response, the government established maximum permissible concentrations under the 1962 Smoke and Soot Control Law (e.g. 2,800 ppm of SO_2 in the stack gases, for petrochemical facilities). This failed to halt the rise in ambient urban SO_2 levels, which was being driven by increasing industrialization. Progressive local authorities began to take matters into their own hands, using their planning powers and moral pressure to persuade owner/operators of new plants to adopt tougher standards.[11]

In response, the government replaced the Soot and Smoke Control Law with the 1968 Air Pollution Control Law. Under the new law, SO_2 levels were controlled by the 'K-value regulation' which permitted higher concentrations of SO_2 from facilities with higher smoke stacks. This 'high stack' policy was similar to the UK approach and was thought to be the only 'end-of-pipe' option available at the time. government-supported research on flue-gas desulphurization had started in 1963, but was not ready for commercial application.

The main tool available to the metropolitan authorities responsible for enforcing the Pollution Control Act was to promote fuel-switching from coal to oil and to set strict standards for the sulphur content of fuel oil. New fuel oil desulphurization techniques allowed them to drive down their local K-values and steadily reduce ambient SO_2 concentrations. As the local authorities demonstrated what could be achieved, the central government progressively reduced the minimum permitted K-value, from 20 in 1969, to 3 in 1976. This, in turn, allowed local authorities to take advantage of new strategies and technologies, as they became available.

Recognizing that a more general urban air quality problem than high SO_2 levels existed, local authorities turned their attention to NOx during the 1970s. Advanced boiler design techniques such as flue-gas recirculation and two-stage combustion have been in use in the Tokyo electric utility's oil-fired power stations since 1973. Low-NOx burners were widely employed in 1977 and the first SCR system was installed in a heavy-oil power unit at Yokosuka power station, in Tokyo Bay, in 1978.[12] By this time, most of Japan's electricity generation was based on oil or gas, but these new control technologies were adapted to the smaller number of coal power stations. Two boilers at Yokosuka running on a coal–oil mixture had flue gas recirculation, two-stage combustion and low-NOx burners by 1977, although SCR was not fitted to these units until 1985.

---

[11] *Outline of Air Pollution Control in Japan*, Air Quality Bureau of the Japan Environment Agency, December 1993.

[12] *Energy and the Environment*, Tokyo Electric Power Company, July 1992, p. 7.

10.3.4. The United States

The Clean Air Act of 1970 established national air quality standards for six pollutants, SO_2 and NOx among them. From this beginning a range of schemes for regulating air quality were developed.

New source performance standards (NSPS) were created by the EPA to restrict emissions from new industrial facilities, including power plants.[13] The 1977 amendment to the Clean Air Act recognized that many urban areas were failing to meet the early goals by creating formal *non-attainment* areas. State authorities covering non-attainment areas were required to establish plans which would bring them into attainment by certain dates.

A legal case had created the concept of *prevention of significant deterioration* (PSD), which Congress integrated into the 1977 CAA (see Chapter 8). The PSD requirement disallows construction of any new pollutant sources, or major modifications to existing sources, if these might cause a significant increase in one of the six controlled pollutants. This applies even to those areas where the air quality standards set by the CAA are being attained.

These health-motivated efforts to control local air quality were successful in eliminating high levels of SO_2 in urban areas. Levels of NOx from fixed sources also fell, but increasing emissions from vehicles are contributing to continuing non-attainment of the national NOx standard in several areas. Overall SO_2 and NOx emissions from power stations fell slightly from the mid-1970s to 1990, despite a near doubling in coal consumption. This was achieved by the gradual replacement of old plant with new plant subject to NSPS emission limits (the SO_2 limit effectively requires FGD on new large power stations).[14] However, total national emissions were projected to rise again with growing demand for electricity.

In the 1980s, political attention shifted from the health impacts of burning coal to the ecological damage caused by acid rain. This was to lead to the

---

[13] United States Environmental Protection Agency, 'The New Clean Air Act: What it means for you', *EPA Journal*, 17, 1, 1991.

[14] James E. Evans, 'Electric utilities and clean air - the progress continues', in *Power Generation Technology: The International Review of Primary Power Production 1992*, Sterling, London, 1991.

novel SO_2 trading programme which is described in detail later in this chapter. Sources of NOx remain subject to NSPS, PSD and the original air quality limits but are free of any national cap on overall emissions.

10.4. Acid Rain in Europe

10.4.1. Origins of Acid Rain as a Political Issue

During the 1960s and 1970s, Scandinavian countries began to complain of the acidification of lakes. Scientific investigations, proving that acidic gases are transported over thousands of kilometres, suggested that emissions from the large European industrial countries were responsible for this acidification. The issue surfaced in international fora such as the OECD, which took up the topic in 1969 at the initiative of Sweden and commenced a broad-based measuring and research programme on long-distance, transboundary air pollution, in 1973. The political profile of acid rain was raised at the first United Nations conference on the environment held in 1972 in Stockholm. Throughout this period, and up to the late 1970s, the response of European Community countries, including Germany, was characterized by a lack of concern for possible damage to ecosystems in distant countries.

German scientists, policy-makers and environmentalists did not show a great deal of interest in acid rain until damage to fir trees was noticed in southern German forests and the findings of the Solling research project were published at the end of the 1970s. This project established a link between airborne pollutants and damage to forest ecosystems of the Solling, a low mountain range in Lower Saxony located relatively far from any power stations.[15]

Forest damage did not strongly enter public awareness until the beginning of the 1980s. A series of articles on forest damage in the weekly news magazine *Der Spiegel*, in November 1981, played a significant role in this process. From then on, forest decline became a daily news subject. The issue tapped into deeply held values of the German people: for most Germans the forested, mountainous areas of their country were the embodiment of nature and they viewed forest decline as the beginning of the wide-scale destruction of nature.

---

[15] OECD, *Country Studies*, p. 186, see note 7.

10.4.2. The Political Response in Germany

Acid rain united several disparate strands in German politics. The Green Party had emerged during the 1970s from local citizen action groups as an extension of the left-wing and anti-authoritarian peace movements of the late 1960s. Initially, the Greens argued for more coal burning as an intermediate step in replacing their main target, nuclear power, with renewable forms of energy. By the end of the 1970s, some regional Green groups were arguing against nuclear *and* coal, and by the time of the 1983 election the national Green agenda was vociferously anti-coal, anti-nuclear and in favour of greater energy efficiency and controls on rampant consumerism.

By 1982, the ruling Social Democrat Party (SPD) was looking tired and tarnished by recession. Since 1969 it had ruled in coalition with the much smaller Free Democratic Party (FDP) and up to the mid-1970s adopted a progressive approach to environmental issues. This changed with the departure of Chancellor Willy Brandt. The SPD shifted back towards its socialist roots in industrial coal mining areas and adopted a strategy of isolating and ridiculing the emerging Greens. Their FDP partners became increasingly uncomfortable with this approach. As a party of the educated, liberal-minded middle class they were head-to-head with the Greens. By 1982 the FDP had decided it would be wiser to adopt the Green agenda which clearly appealed to its traditional electorate. The SPD made a belated attempt to accommodate the FDP by accepting a long-delayed air pollution control law (the Large Firing Installations Ordinance, or GFAVo, described in detail below) but this was widely seen as too little, too late. A general public mood of doom and anger over other issues, such as the government's acceptance of the Pershing nuclear missile, led to the FDP's decision to abandon the coalition in October 1992.

The FDP then entered into coalition with the Christian Democratic Union (CDU) and its sister party, the Bavarian Christian Social Union (CSU). Chancellor Kohl's CDU had a history of strong support for nuclear power. Fears of forest decline had shifted public opinion back towards nuclear power to some extent, making this policy less of an electoral liability. The CSU, which operates solely in the state of Bavaria, also supported nuclear power and had already embraced tight controls on power station emissions, partly as a political weapon against the SPD's industrial heartlands which were now believed to be destroying 'Bavaria's' forests.

Following the March 1983 election, which confirmed the CDU/CSU/FDP coalition and ushered the Green Party into the Bundesrat for the first time, the GFAVo was quickly finalized and implemented, and became law in June 1983. During these negotiations the Green Party was able to wield its share of power in the Bundesrat directly. It alerted members to a government proposal that the GFAVo should not apply to individual boilers with power output of less than 50MW. The expectation had been that the threshold would apply to individual *sites*, many of which contained several small boilers. This proposal, which would have seriously reduced the applicability of the GFAVo, was rejected by the Bundesrat.[16] Also during this period, the Council of Environmental Advisors was responsible for a change in the draft GFAVo which replaced requirements for specific types of emission control equipment with emissions limits, leaving utilities free to choose the best methods of emission control.

10.4.3. German Acid Rain Legislation

Origins of the GFAVo During the 1970s, German electricity utilities were concerned at the tendency for some states to insist on stringent emission controls on power stations. Federal guidance on the state of the art in air pollution control technology was contained in the 'Technical instructions on air quality control', or 'TA Luft'. This set minimum emission levels for a wide range of industrial activities but gave state authorities discretion to go beyond these limits. Occasionally this led to legal challenges, particularly when states demanded retrospective action which industry felt was not 'economically defensible'. To clear up the confusions in the power sector, where some authorities would demand late changes to engineering specifications for new plants, some utilities began to press for a binding regulation on power station emissions. This would remove the states' discretion to set higher standards and would be advantageous to the utilities if the standards were relatively lax.

Against this background, the Federal Environment Agency (UBA) circulated a draft regulation proposing SO_2 limits for large combustion plants to interested branches of industry, in 1980.

---

[16] Boehmer-Christiansen and Skea, p. 197, see note 3.

Negotiating GFAVo: The Advantages of Secrecy Although UBA presented the initial draft for the GFAVo, it was the Industry Ministry (BMWi), with responsibility for energy policy, which took charge of subsequent negotiations. Officials at BMWi took stock of the concerns of the large combustion plant users – electricity utilities and large industrial facilities such as chemical plants – over the costs of reducing emissions. At the same time they needed to know the current and potential effectiveness and costs of control technologies. These technologies were developed and supplied to the users by a small number of highly specialized engineering firms, which feared a souring of relations with their customers, if they were seen to provide information which would lead to stringent and costly regulations. To get around this, the BMWi consulted openly with the major energy users, but met in secret with the technology suppliers.

From Japanese experience with FGD, and their own R&D efforts, suppliers such as Lurgi informed the BMWi that a 95% reduction in SO_2 emissions would be possible. The ministry accepted that a 90% reduction could certainly be achieved but considered the 95% target to be dependent on further development.

This dual-track negotiation process also gave rise to the attempt, late in the implementation of what was to become the GFAVo, to apply the 50MW cut-off for application of the regulation to individual boilers, not sites. This was perfectly reasonable in the eyes of the utilities, which pointed out that installing control technology on boilers below a certain size would be hopelessly uneconomic. As mentioned earlier, this change was defeated by the Greens in the Bundesrat, forcing many small coal-fired combustion plants to be closed down, or switched to other fuels, when the legislation took effect.

In the case of NOx reduction, it was not clear that the emission reductions being achieved by the Japanese would be immediately achievable in Germany. A wider range of coal was used in Germany and a much larger proportion of plant was run below peak capacity, with more frequent stops and starts. More importantly, much of the German plant was of the wet bottom boiler type.[17] This type of boiler has intrinsically higher rates of NOx formation than dry bottom boilers, which predominate in Japan. A further

[17] In a wet bottom boiler combustion takes place at high temperature, above the melting point of ash, which is deposited at the bottom of the boiler in molten, or wet, form.

Table 10.1 Emission Limits for Coal-fired Power Stations in the Federal Republic of Germany under the 1983 Large Firing Installations Ordinance (GFAVo)

Thermal Power	SO_2 (mg/m³)	NOx (mg/m³)
< 50 MW	exempt	exempt
50–300 MW	2,000	800
> 300 MW	400/650[1]	800

[1] For all plants over 300MW sulphur dioxide must be reduced by at least 85%. In addition, for fuels with very high sulphur content the highest allowable concentration is 650 mg/m³ and for all other fuels 400 mg/m³.
Source: Ordinance on Large Firing Installations, 13th Ordinance Implementing the Federal Imission Control Law, 22 June 1983, Articles 5 and 6.

confusing factor was the very wide range of NOx emissions recorded in plants with no emission controls. One review of emissions from dry bottom boilers in Germany, the United States and the United Kingdom, cites a range of 800–2,150 mg/m³, from several sources.[18]

Faced with these technical uncertainties, and under intense public pressure to act quickly, the German government settled on 800 mg/m³, as a level of NOx emissions which should be achievable for most large boilers through basic combustion modifications.

Form and Function of the GFAVo The emission limits for SO_2 and NOx established by the 1983 GFAVo are shown in Table 10.1.

Implementing the GFAVo Utilities and other combustion plant operators had one year from the date the ordinance was adopted to publish their plans for compliance. This sparked off a scramble among process technology firms to come up with their own implementations of the known control technologies, particularly FGD. The rewards for firms with good control systems seemed enormous. Estimates made at the time, that the cost of complying with the regulations would be in the region of 20–30 billion DM, have been proved correct. The federal government encouraged

[18] Anna-Karin Hjalmarsson, *NOx Control Technologies for Coal Combustion*, IEA Coal Research, IEACR/24, June 1990, p. 31.

experimentation with new processes by providing utilities with a subsidy of 50% of the costs of their investments in FGD plant.

Officially, the GFAVo brought national consistency to emission standards by removing the discretion previously enjoyed by state authorities. In practice, the nature of the commercial relationships between utilities and the technology suppliers exerted a further pressure on emission levels.

The utilities were the source of this pressure through their practice of demanding performance guarantees from the engineering design firms which were contracted to supply their FGD installations. Suppliers were forced to squeeze maximum performance from the processes they were developing, in an attempt to provide a margin of error which could absorb poorer than expected performance in any part of the system.

The risks for some of these firms were enormous. In some cases, the processes they were attempting to implement had not been tested much beyond laboratory scale. For these firms, the downside risk could greatly outweigh the upside. Their problem stemmed from their commercial role in a typical FGD contract. Most of the value of a contract is bound up in straightforward construction work which was usually sub-contracted to construction firms. The technology supplier would typically retain only around 10% of the contract value. One engineering design firm, Walter & Co., supplied an FGD retrofit based on a system which it had not fully demonstrated at pilot scale. When the plant was commissioned, it produced a highly visible plume which was unacceptable to the local authorities. The customer forced Walter & Co. to remove the FGD plant at its own expense (a construction task equivalent to installing the original plant), forcing it into bankruptcy.

Because of their fear of this eventuality, the margins of error which process designers built into their designs for FGD plant were substantial. In most cases the original design criteria were met, and were often greatly improved upon. In 1993, the German Industry Ministry was aware of recent plants fired with low sulphur content hard coal which were operating with SO_2 emissions of less than $100mg/m^3$, compared to a legal requirement of $400mg/m^3$.

The GFAVo also failed to eliminate entirely the influence of local regulatory authorities on emission limits. State regulators have retained increased powers in areas which they designate as having poor ambient air quality, and operate a form of total emissions control in these areas. On occasion they have demanded very stringent emission limits when giving approval for new coal-fired plant. Word of these 'latest' state-of-the-art emission limits is then communicated, in a somewhat haphazard manner, among state officials, local citizen action groups and Green groups. When utilities make applications for approval for new plants elsewhere, in areas with no special planning considerations, local objectors will often be aware of the standards being applied to those stringently controlled plants. To defuse such opposition during mandatory public consultation, utilities often feel it is prudent to propose the latest emission standards from the outset.

This ratcheting effect is similar to the process which occurs in Japan, but with two major differences. First, in Japan a *formal* process for advising all regulators of the latest state of the art has been put in place and, second, this information is based on *actual* performance rather than the performance target contained in the planning consent. As a result, the Japanese 'ratchet' tends to be more powerful and consistent.

10.4.4. The Export Strategy Myth

When German politicians adopted the GFAVo, they did so almost entirely on environmental grounds and in response to overwhelming pressure of public opinion. However, the perception in some other countries was quite different, and this misunderstanding of the origins of the German legislation was to colour later international negotiations on acid rain.

The United Kingdom, especially, was mistrustful of German intentions. In the early 1980s confidence in the United Kingdom was at a low ebb, following the budgetary embarrassments of the late 1970s under the last Labour government and the industrial disruption of the Conservative government's soon-to-be-abandoned monetarist policies.

Germany's about-face on acid rain was widely misunderstood in the United Kingdom. Observers there felt that Germany had acted on the basis of a very poor scientific link between forest damage and acid rain. Yet their FGD retrofit programme would be hugely expensive. Surely no rational government would take this course? Policy-makers in the United Kingdom looked for alternative explanations for the German acid rain policy. One view held that, since Germany was rich and successful, its utilities and industries could easily afford an expensive acid rain control programme. By forcing other countries to do the same, they would impose the same costs on weaker competitors, debilitating them even further. In reality, Germany's own confidence was at a low ebb in 1983 and German industry feared a loss of competitiveness as acid rain controls pushed up electricity prices. The government, having chosen to reduce emissions for environmental reasons, wanted to level the playing field by requiring similar measures elsewhere. This made sense on environmental grounds as well, since efforts confined to Germany were likely to be overturned if emissions continued to grow in nearby countries.

Another view of German acid rain policy was to interpret it as a deliberate strategy to create a demand for German emission control equipment and expertise in other European countries. This cannot be a complete explanation for Germany's 1983 decision to adopt the GFAVo and to internationalize acid rain controls. However, this view gained support elsewhere as the protracted international negotiations dragged on through the 1980s and Germany, in order to meet the requirements of the GFAVo, inevitably developed a strong emission control industry. Indeed, once Germany had committed itself to major domestic efforts, a desire to make some economic return on this investment probably formed an element of its international negotiating strategy.

10.4.5. A Regional Approach: UNECE and the European Community

As part of its swift conversion to action on acid rain, with the change of government in 1982–3, Germany sought to pursue acid rain controls internationally. The natural forum for this effort was first the United Nations Economic Commission for Europe (UNECE), and later the European Community.

UNECE Negotiations The UNECE is the European regional commission of the United Nations. Its members are the European countries, including the former Soviet bloc countries, plus Canada and the United States. Following a Soviet political initiative, the UNECE countries adopted a convention on Long Range Transboundary Air Pollution (LRTAP) in 1979.

Signatories committed themselves to limiting, and if possible reducing, air pollution by using best available technology where economically justified. Scandinavian countries had pressed for at least a standstill commitment on SO_2 emissions. Several countries, including the United States, the United Kingdom and West Germany blocked this proposal.[19] The LRTAP had no fixed targets or time scales, relying instead on its potential as a source of diplomatic embarrassment to pressurize signatories into taking action.

At the Stockholm Conference on Acidification of the Environment in 1982, Hans Dietrich Genscher, West Germany's Foreign Minister and leading light of the FDP, revealed Germany's change of heart when he supported the Scandinavians' call for SO_2 reductions to 30% below 1980 levels by 1993. Most Soviet bloc countries supported the proposal – with the proviso that the reduction applied only to the level of emissions they 'exported' to other countries – as did many other European countries and Canada. This proposal was translated into a protocol to the LRTAP Convention, the 'Helsinki Protocol', and the signatory countries became known as the '30% club'. The United States and the United Kingdom could not be persuaded to join. It was now clear to German policy-makers that they would need to tackle acid rain through the European Community, where mandatory Directives were a possibility, and horse-trading across a range of issues might persuade reluctant member states to modify their positions.

Towards a European Community Directive German negotiators, aided unwittingly by European Commission officials, got off on the wrong foot when they first introduced the idea of binding acid rain controls in the EC. By giving the impression of forcing German domestic legislation on other countries, they raised deep and lasting suspicions, especially in the United Kingdom, that their real agenda was to export first their regulations and then their pollution control technology.

In June 1982, Germany issued a memorandum to the European Council on the need for EC action on acid rain. A year later, under the German presidency, the Council discussed the environment for the first time and called for action on acid rain. Working closely with the Germans, European Commission

---

[19] Elsom, pp. 320-22, see note 2.

officials produced in December 1983 a draft directive on emissions from large combustion plants (LCP Directive).

The draft LCP directive was modelled on the German GFAVo. For new plants, similar emission limits were proposed for plants of similar power burning similar fuels. As in Germany, FGD would be essential on new plants above 300 MW thermal power. In addition to plant-based emission limits, the draft directive also proposed that each member state should achieve uniform reductions in overall emissions from large combustion plants. By 1995, SO_2 was to be reduced by 60% and NOx to be reduced by 40%. This proposal implied a massive retrofit programme for existing coal-fired power plants throughout the EC, along the lines of the programme Germany was pursuing domestically.

During the protracted negotiations, which began in 1984, German officials adopted a rigid approach, even on apparently minor details of the proposal.[20] The combination of clear German origins and subsequent German rigidity on technical issues fed the UK officials' suspicions that Germany's objectives were as much part of a long-term export strategy as they were environmental.

Final agreement on the LCP Directive was not achieved until 1988.[21] This sets an EC-wide target of a 58% reduction from 1980 SO_2 emissions, by 2003. Emissions of NOx are to be reduced by 30%, by 1998. Targets vary from one country to another (see Tables 10.2 and 10.3), with some of the poorer member states even securing agreement to increase their emissions, offset by greater efforts in other countries.

For new plants, the limits in the Directive are remarkably similar to those in the 1983 German GFAVo. The main changes are a sliding scale for SO_2 emissions, from 2000mg/m^3 for 100MW plant, to 400mg/m^3 for plants of 500MW and above. Details of the SO_2 and NOx limits are given in Table 10.4.

The LRTAP 1st Review In 1994, the Helsinki Protocol to the UNECE convention was updated. In the process several of the countries which refused to join the original '30% club' were brought on board. Most notable, perhaps, was the United Kingdom.

---

[20] Boehmer-Christiansen and Skea, p. 249, see note 3.

[21] 88/609/EEC, *Official Journal* of the European Communities L336, 7 December 1988.

Table 10.2: SO_2 Emission Reductions, EC Large Combustion Plants Directive (% reduction over 1980)

	1993	1998	2003
Belgium	–40	–60	–70
Denmark	–34	–56	–67
France	–40	–60	–70
Germany	–40	–60	–70
Greece	+6	+6	+6
Ireland	+25	+25	+25
Italy	–27	–39	–63
Luxembourg	–40	–50	–60
Netherlands	–40	–60	–70
Portugal	+102	+135	+79
Spain	0	+24	–37
United Kingdom	–20	–40	–60
EC 12	–23	–42	–58

Table 10.3: NOx Emission Reductions, EC Large Combustion Plants Directive (% reduction over 1980)

	1993	1998
Belgium	–20	–40
Denmark	–3	–35
France	–20	–40
Germany	–20	–40
Greece	+94	+94
Ireland	+79	+79
Italy	–2	–26
Luxembourg	–20	–40
Netherlands	–20	–40
Portugal	+157	+178
Spain	+1	–24
United Kingdom	–15	–30
EC 12	–10	–30

Table 10.4: Emission Limits for New Coal-fired Power Stations, EC Large Combustion Plants Directive

Thermal Power	SO_2 (mg/m³)	NOx (mg/m³)
100–500 MW	2,000–400	650/1,300[1]
> 500 MW[2]	400	

[1] For wet-bottom boilers burning coal with a volatiles content of less than 10% the limit is 1,300mg/m³.
[2] For high sulphur content indigenous coal only 90% SO_2 removal is required.

During the negotiations on the updated Helsinki Protocol, the 'critical loads' approach dictated the levels of SO_2 emission reductions being sought by the progressive countries. A critical loads analysis of ecosystems, e.g. forests or rivers, is an attempt to determine the minimum rate of acid deposition at which damage occurs. From among several modelling groups working on critical loads, the UNECE chose to base its proposed emission reductions on the RAINS model developed by the International Institute for Applied Systems Analysis (IIASA) based near Vienna.

According to the IIASA model, the United Kingdom's contribution to critical loads implied that a reduction of 88–92% in its emissions, relative to 1980, would be necessary in order to protect the most sensitive ecosystems. The original proposal for the revised SO_2 protocol called for this to be achieved by 2000. Several countries, including the United Kingdom, objected to the scale and rapidity of the reduction. Eventually a compromise solution emerged, based on reducing by 60% the gap between existing commitments and the targets from the original modelling. For the United Kingdom, this implied a 79% overall reduction on 1980 emissions. The parties finally agreed on a target date of 2010 and the United Kingdom signed up to reductions of 50% by 2000, 70% by 2005 and 80% by 2010.

This was in keeping with United Kingdom projections of a substantial further decrease in coal utilization, and was expected to be achieved without any significant further retro-fitting of desulphurization units. In contrast to the earlier negotiations on the first protocol and on the EC Directive, the UK government had no reason to fear that German technology suppliers might profit from the new protocol.

During the negotiations, the United Kingdom's willingness to go further than the UNECE proposal, albeit by only 1% and over a longer time scale, deflected a lot of the criticism towards other countries which were proving more reluctant, notably Spain, Portugal and Denmark.

The critical loads method of analysis was itself a source of controversy. The US Environmental Protection Agency considered the technique to be too poorly established to form the basis of policy-making. This assessment, and its unwillingness to revise its new SO_2 emission permit trading system, prevented it once again from signing the sulphur protocol. One major objection to the critical loads analysis was its treatment of very low levels of acid deposition. Many of the critical loads calculated for vulnerable ecosystems, showing up as a discrete area on the maps produced by the modellers, were lower than the natural background level of acid deposition at these sites.

Something was clearly amiss if the model was predicting that harm would occur at levels of deposition which were normal even before the industrial age. In the end this anomaly was never fully resolved, although it is likely to be a major issue of contention if, at the next review of the Protocol in 1996–7, the critical load models again lead to calls for emission reductions to go even further.

10.5. Impacts on Technologies

Close examination of the evolution of the major NOx and SO_2 control technologies lends strong support to the view that technological innovation is driven by needs. It follows from this, that the market structure and level of demand are the dominant factors in determining where ownership of the latest generation of technologies resides.

When the Americans, and then the Japanese, became concerned about local air quality problems due to SO_2 emissions, they went to London to examine the early FGD plants at Battersea and Bankside. In the United States, Babcock and Wilcox was the first firm to commercialize the wet limestone FGD process, in response to the demands of the 1970 Clean Air Act. Soon after this, the process was licensed to Babcock-Hitachi of Japan as demand for FGD there grew.

Licensing was essential because the power equipment market has traditionally been a closed one, where public or quasi-public electricity utilities maintain exclusive relationships with a small number of national heavy engineering firms. Where local firms have the potential to supply equipment under licence, foreign competitors have generally found that this is the only means of participating in the market.[22] This has remained the case until recently, when conscious efforts have been made to open up trade in 'strategic' industries.

In Europe, the common market should theoretically have created a free, competitive market in power equipment a long time ago. In practice, utilities flouted free trade with impunity and many continued to purchase almost exclusively from local firms. Even the EC Directives of the early 1990s on public utility and government procurement have not fully rectified this situation, although they have exposed tendering processes more fully to public scrutiny and to greater likelihood of challenges arising from firms which feel they have been treated unfairly. Globally, the increasing dominion of the General Agreement on Tariffs and Trade has forced some changes in behaviour by utilities, although complaints by would-be suppliers of favouritism to local firms persist, whether it be from European firms trying to sell to the United States, or US firms trying to sell to anyone.

Against this background of closed markets, it is difficult for licensers to maintain their grip on the basic technologies and processes. In Japan, Hitachi soon escaped from its licensing arrangements by modifying the FGD processes. This was driven as much by the particular characteristics of the Japanese market as by a desire to stop paying licence fees. In the United States, large power stations had ample space to build lagoons to store the calcium sulphite slurry produced by the Babcock and Wilcox process. In Japan, power stations tended to be located in more densely populated areas and many FGD plants were also required for other types of industrial site which often coexisted with residential areas.

[22] As a last resort, utilities have bought equipment directly. For example, the Tokyo electric utility bought gas turbines from GEC of the United States for new LNG-fired power stations in the early 1980s, when Japanese firms had no previous experience in this area. Recent purchases have been from Japanese firms which moved into the market when they saw there would be substantial local demand.

Babcock-Hitachi substantially modified the FGD process to suit the Japanese market. Now the end product of desulphurization was calcium sulphate (gypsum) which could be sold to cement manufacturers, thus avoiding any need to store large volumes of waste.

When Germany began its massive desulphurization programme in the early 1980s, German firms (including another remnant of the original Babcock group – Deutsche Babcock) licensed the Babcock-Hitachi process. Once again the process was modified for the home market. Responding to the fact that German utilities were more cost-sensitive than those in Japan, the German firms simplified the FGD processes and in doing so changed them sufficiently to escape the licence arrangements.

Babcock Energy in the United Kingdom agreed licensing terms with Babcock-Hitachi for selling the Babcock-Hitachi process in Europe. Similarly, John Brown of the United Kingdom licensed a process from General Electric in the United States, which also licensed its process to local firms in other European countries and prevented licensees from competing in one another's markets. The United Kingdom firms suffered from a lack of any domestic orders in the period when the CEGB was dismissive of FGD. With no sales there was no possibility of adapting the technology and perhaps escaping the licences, as had occurred earlier in Japan. With no domestic market to speak of, there was no real possibility of competing in foreign markets.

When in 1986 the CEGB finally accepted that Scandinavian lakes were being damaged by acid rain, it decided to retrofit FGD to Drax, its latest large coal-fired power station. The contract for this work went to Babcock Power, which licensed the FGD process from Babcock-Hitachi.

Licensing, followed by local adaptation, was as important to commercial success with SCR systems for controlling NOx emissions as it was with FGD. Germany's GFAVo regulation effectively required existing plants to install SCR. The catalyst technology for SCR had been developed in Japan in the 1970s and was initially licensed to German manufacturers. But soon the technology was modified (for example, the grid spacing of the catalyst's cells was changed) and now German firms own the technology and operate their own catalyst manufacturing plants.

The United Kingdom has, perhaps surprisingly, established a strong position in the other main NOx control technology, low-NOx burners (LNBs).

International Combustion, a long established UK engineering firm, licensed LNB designs from Combustion Engineering which was once the largest supplier of boiler equipment in the United States. International Combustion quickly developed and improved the technology, making it its own. In the early 1990s, International Combustion had the largest share of the American market for LNBs. Meanwhile, Combustion Engineering suffered badly during the 1980s. The announcement and then long delay of a major amendment to the Clean Air Act caused a climate of uncertainty which dissuaded American utilities from ordering new coal power stations during the 1980s. Having seen its business collapse to almost nothing, Combustion Engineering was bought by the Swedish–Swiss-owned ABB.

It is certain that closed markets for power generation equipment have encouraged domestic firms to modify the processes they license from abroad, and make them their own. However, it is likely that this would have happened even if, for example, Babcock and Wilcox had been able (and willing) to operate independently in Japan, or Hitachi had been given the same opportunity in Germany. Most of the emission control technologies are complex, multi-process installations which often need to be adapted to suit each plant and other domestic market conditions. There is plenty of scope for slipping out of patent restrictions through major or minor modifications, and local firms are likely to attempt this wherever they see a significant potential market.

This process is likely to repeat itself in areas of rapid economic growth, such as China and India. Many Western technology suppliers are gleefully aware of the business prospects in these areas, once their governments become serious about tackling emissions. However, these markets are large enough to provide plenty of scope for further modification of all the major emission reduction technologies and, therefore, plenty of scope for domestic firms to take over the technological lead.

None of this should be interpreted as a recommendation to Western firms not to pursue these markets. There will undoubtedly be plenty of opportunities to earn healthy profits in developing countries. However, the history of these technologies demonstrates that a large domestic regulation-driven market will give firms a technological lead and overseas success. But when the action shifts elsewhere, so does the technological lead, and firms based in the growing markets will become the world market leaders, on the back of their domestic markets.

A business of this nature cannot possibly justify strengthening domestic regulations just to capture overseas sales – long-term dominance of world markets is an unrealistic hope. Close examination of the regulatory history reveals that this has never been attempted. Governments set environmental objectives for environmental reasons.

10.6. Marketing Acid Rain

In the United States, the controversy in Europe and complaints from Canada of lake acidification caused by US power plants inspired the influential National Academy of Sciences to pronounce that acid rain was a potential problem. In response, the government set up the National Acid Precipitation Assessment Program (NAPAP) to examine in greater detail the scientific evidence for acid rain damage to forests and lakes. As evidence of acidification of lakes in the United States began to emerge during the mid-1980s (15–25% of lakes in the Adirondack Mountains are classed as being 'chronically acidic'), public concern over acid rain began to mount. The results of the NAPAP study formed one of the main sources of evidence for the scale of SO_2 reductions required by the Clean Air Act Amendments of 1990.[23]

These amendments authorized the EPA to create a revolutionary method of controlling SO_2: the tradable emission permit (TEP). This bold experiment in using market instruments may have profound impacts on the use of such instruments elsewhere in the world. Its effects on technological development and innovation are complex and may not be fully understood for another decade.

10.6.1. Origins of the SO_2 Tradable Emission Permit Scheme

During the early to mid 1980s, while NAPAP was building a scientific consensus on the causes and effects of acid rain, a group of American economists were laying the theoretical framework for a new instrument to

[23] *EPA Journal*, US Environmental Protection Agency, January–February 1991.

control pollution. They proposed that regulators should create a market for pollution by dividing the total emissions they could allow each year from all sources into a large number of permits to pollute. Firms could buy and sell the permits as required at prices determined solely by supply and demand, and would be free to pursue any emission reduction strategies which best suited their circumstances. In perfect trading conditions, the level of emissions required by the regulator would be achieved at the minimum cost predicted by economic theory.

In the lead-up to the 1990 CAA Amendments, influential economists and utility executives, who had been convinced of the potential cost savings, were pressing for the administration to tackle acid rain, and other issues, through a TEP scheme. The Environmental Defense Fund submitted a proposal to the EPA for just such a scheme, which would deliver the 10 million tonne annual cut in SO_2 emissions that the environmental lobby had rallied around.

The EPA's initial reaction was negative, because of the uninspiring outcomes of earlier pollution credit schemes. However, it was won over by a combination of factors. First, it quickly realized that the proposed scheme was closer to a genuine implementation of the economists' theoretical permit market, whereas earlier schemes had been limited to giving credit to firms which proved they had gone beyond previously mandated controls. Second, they believed that utilities would accept a greater overall reduction with a TEP scheme than with traditional regulation. Third, the scheme would introduce a permanent, nationwide cap on a pollutant for the first time in the United States. Finally, the market approach would appeal to the conservative Bush administration and accord with the prevailing rhetoric on lightening the regulatory burden on industry. The approach duly gained President Bush's support and formed part of the proposed Clean Air Act Amendments submitted to Congress in 1989.

In Congress, every part of the bill was, as usual, subject to intense lobbying by industrialists and environmentalists, but, according to observers in the EPA, the trading scheme proved to have an unexpected, natural ability to survive which allowed it to sail through this treacherous process and emerge basically unchanged at the end. In political science terms, the TEP scheme created a new means for political agents – Congressmen, administration officials, and lobbyists – to exercise their institutional power, and distracted them from the normal expression of that power. This is best illustrated by the battle over the means of allocating permits to the midwestern states.

Delegates from these states, where high sulphur coal is produced and burned, were concerned about the cost to their regions of controlling emissions – costs which would show up in consumers' (i.e. voters') electricity bills. If the proposed regulation had been of a traditional variety, requiring named power stations to reduce emissions to set levels, the political fighting would have focused on the required level of emissions reductions and, by extension, the robustness of the science of acid rain. This would have forced all of the industry lobbyists and their Congressional representatives to join with the midwestern constituency in a traditional industry–environment confrontation.

Instead, all sides seemed mesmerized by the new choices and opportunities for negotiation opened up by the new regulatory toy. The delegates from the midwestern states put their efforts into obtaining favourable treatment and argued for extra permits to be allocated to their local utilities. This matter was resolved in their favour by the creation of a raft of special categories, on flimsy pretexts, which provided sufficient cover for the politicians to deviate slightly from the strictly objective allocation proposed by EPA officials.

10.6.2. Design of the TEP Scheme

Shortly before the 1990 CAA Amendments were passed by Congress, the EPA established a consultative body of interested parties, to assist in drawing up the acid rain strategy. The Acid Rain Advisory Committee consisted of 44 members drawn from utilities, fuel suppliers, state environment and utility regulators, environmental groups, academics and EPA officials, and first met just one month after the bill was passed. This represented a new style of working within the EPA. Previously, the 'rule-making' process following approval of an environmental bill had been regarded as an internal agency responsibility, leading to a draft implementation plan. Upon revealing the plan to the public the EPA would normally announce a formal consultation period, typically of 30–60 days.

The new approach embodied in the Acid Rain Advisory Committee was a conscious effort for a more open, inclusive rule-making process. This involved a series of public meetings with local groups throughout the country, detailed work on key aspects of the acid rain strategy in sub-committees of the Acid Rain Advisory Committee and an 'open door' policy among EPA officials. EPA officials estimate that this doubled the effort required to draft the implementation plan but feel it was worthwhile as a means of minimizing both the unexpected objections which derive from the traditional rule-making process, and the threats of court action which often arise when outside parties believe that the EPA proposal does not respect the underlying legislation.

Much of the framework of the TEP programme described below was already determined by the primary legislation passed by Congress. The trading scheme applies only to electric utilities, which are responsible for the great majority of SO_2 emissions. Their overall output of 18 million tonnes in the 1985 baseline year will be reduced to 8.9 million tonnes per annum, at some date after 2010 – a reduction of 10 million tonnes per annum from the business-as-usual scenario.

This will be achieved in two phases. In Phase 1, beginning in 1995, 110 power plants with capacity greater than 100 MW and emissions over 2.5 lbs per million British Thermal Units (mmBTU) will be required to reduce their emissions to this level in the aggregate, by making overall reductions of 3–4 million tonnes per annum. Permits based on this level of emissions have been distributed to each plant. For the most part these plants will easily exceed their reduction targets by retrofitting SO_2 scrubbers. Many will have surplus TEPs which they can trade with other firms, or bank for future use when the more stringent Phase 2 regime takes effect.

Phase 2, beginning in 2000, affects all electric utility power plants with capacity greater than 25 MW (approximately 1,000 plants). Permits for 8.95 million tonnes of SO_2 annually will be allocated to the utilities, based on a 1.2lbs/mmBTU target. Any new SO_2 utility sources must purchase sufficient permits in the market, or find offsetting reductions from elsewhere in the utility. One of the main features utilities and regulators expect to see is a smooth decline in emissions between 1995 and 2010 as the Phase 1 plants exceed their aggregate target and bank permits up to 2000, and these banked permits are used to ease attainment of the Phase 2 standards after 2000 (see Figure 10.1).

Industrial (non-power) sources were excluded from the scheme because they are a small source of SO_2 in comparison with utilities and because they

Figure 10.1 Utility SO_2 Emission Reductions

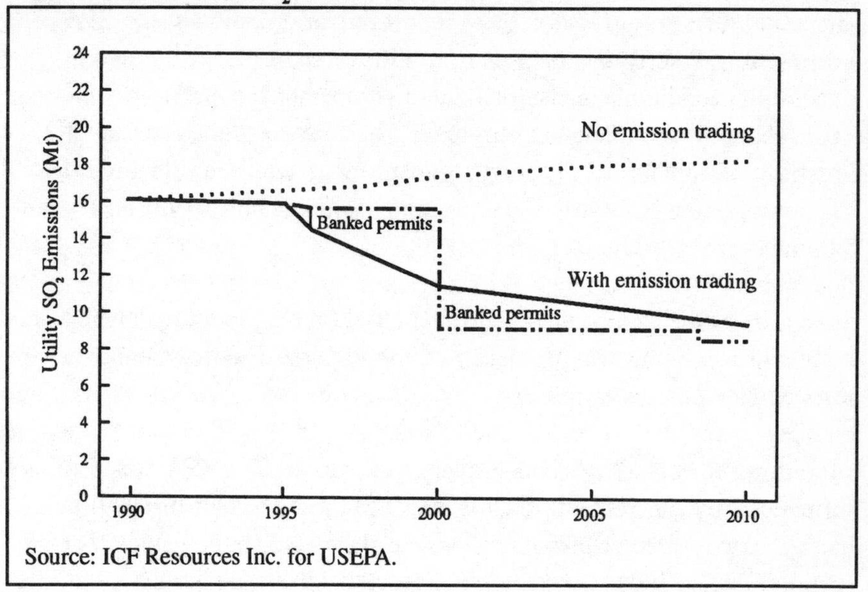

Source: ICF Resources Inc. for USEPA.

would be far more difficult to regulate and monitor. However, an industrial source can 'opt-in' to the TEP scheme by installing SO_2 control measures and receiving corresponding TEPs in return. Theoretically, this should encourage industry to make reductions which are cheaper than prevailing measures in utilities, as they should be able to sell their permits to the utilities at a profit.

Transaction Costs Inevitably, the TEP scheme which has emerged reflects the need to make a transition from a long-established culture of direct, plant-by-plant regulation to a market scheme. This is the reason for the separation into Phase 1 and Phase 2, with Phase 1 being essentially a catch-up exercise for those plants which went unregulated in the past. This hybrid approach has greatly increased the administrative burden of the programme. For example, many power plants have individual boiler units which come under Phase 1, while other boilers on the same site will not be affected until Phase 2. Utilities have an incentive to reduce their use of the Phase 1 boilers and increase generation levels of the Phase 2 boilers. The EPA has felt compelled to respond by requiring the Phase 2 boilers to keep a record of their emissions and generation during Phase 1 and report this information to the EPA. The information is used to reduce permit allocations in line with any shift in generation, and involves just the kind of cumbersome paperwork which the TEP scheme was intended to eliminate. The EPA concludes that the hybrid approach of combining an economic instrument with traditional regulatory 'command-and-control', has created unnecessary complexity and administrative burdens.[24]

Monitoring Emissions One major battlefield throughout the development of the TEP scheme was the issue of monitoring. Environmental groups insisted that all sources must be continuously monitored, while the utilities baulked at the cost of this and favoured the more traditional method of submitting to a stack test once every year or so. The EPA was strongly influenced by the recognition that the utilities now had direct financial incentives to understate their emissions, i.e. they would build a store of unused permits which could then be sold. It foresaw that if a small number of utilities were suspected of false reporting, others would follow suit in a domino effect, and the market would swiftly break down. In the end, all sides agreed that confidence in the market was paramount, and all participating sources were required to install continuous emission monitors (CEMs) in their stacks.

The CEMs automatically record the level of emissions at hourly intervals. This information is electronically reported to the EPA every three months and fulfils a role analogous to the delivery of a cargo in a commodity trading system. The EPA reduces the balance of permits in the utility's account by the amount of emissions reported by the CEM. The EPA estimates that the annualized cost of the CEMs is $180 million – a small fraction of their estimate of the cost savings of the TEP system, although a study by the Electric Power Research Institute put the annual CEM cost higher, at $400 million.[25]

---

[24] Renee Rico, 'United States' Experience in Designing and Implementing an Emission Trading System for Sulphur Dioxide', Acid Rain Division, U.S. Environmental Protection Agency, October 1993.

[25] Reported in *Compliance Strategies Review*, 17 January 1994.

Enforcement Enforcement actions against polluting firms in the United States frequently end up in court. This is largely the result of inadequate monitoring data. With the CEM system, this failing is, theoretically, overcome. The 1990 Clean Air Act recognized this by creating two features not seen before in US environmental legislation. First, penalties are automatic, so the EPA does not need to build a case before imposing a penalty. Second, the EPA applies two distinct penalties to each infringement: a $2,000 fine for each tonne of emissions which is not covered by a permit and forfeiture of permits equal to the tonnage of the infringement. When the bill was enacted in 1990, $2,000 was expected to be roughly three times the market price of a permit, so the total sanction would be four times the cost which a utility might attempt to evade by not buying sufficient permits.

Trading The methods by which permits find their way onto the market is obviously a crucial determinant of the system's success. Impediments to trading such as excessive trading costs, frustrations and delays in making trades, lack of confidence in the veracity of permits on offer, or simply utilities' reluctance to engage in an unfamiliar activity would all make the market illiquid, and prevent the realization of the expected cost savings for utilities having flexibility to choose their emission levels. The TEP programme recognizes this danger and employs the device of an annual forced auction. This includes a 'spot' auction of approximately 3% of the permits allocated to utilities in Phase 1, and a 'forward' auction of permits for Phase 2 which cannot be exercised until the year 2000. The auction provides liquidity, helps to establish a price signal, and ensures that new entrants have an opportunity to purchase the permits they need. The proceeds of the spot auction are returned to the utilities to which the permits were originally allocated.

The main guarantor of the trading mechanism, however, is intended to be a permit tracking system run by the EPA, which would provide a continuous means for purchase and sale of permits.

10.6.3. The Market in Action

Trading has been relatively slow in the run-up to Phase 1. The general picture has been one of substantial transfers between plants belonging to the same utility, light trading between separate firms, but a proliferation of different types of organizations becoming involved, as entrepreneurs from many aspects of the energy industry explore the possibilities of this unique scheme.

Trading Volume The Electric Power Research Institute estimated that industry-wide cost-savings would be maximized with an inter-utility trading volume of over 2 million tonnes per year (the 'perfect trading' volume). However, it believes that the potential supply is likely to be around 1 million tonnes, and that the actual level of trading, based on expectations about utilities' strategic behaviour, is likely to be in the range 500,000 to 800,000 tonnes.[26] This would produce savings of $450 to $850 million per year, on top of annual savings of $1.7 billion due to intra-utility trading, relative to a command and control system.

The first forced auction in March 1993 suffered from the manner in which it was conducted. Bids were invited in advance, but since the EPA did not specify a minimum, the bid prices ranged from 1 cent to $450. On the auction day, minimum prices of $131 and $122 were laid down for Phase 1 and 2 permits respectively. Of the 145,000 Phase 1 permits on offer, only 50,000 were sold, although nearly 80% of the Phase 2 permits were matched with sufficiently high bids. A further peculiarity was that even the highest bidders were held to the bids they had submitted in advance. These were made in the absence of any feedback on the strength of demand, as one would gain from a live auction. In retrospect, it was clear that the forced auction had failed in one of its main objectives: providing a clear price signal.

Market Players On the other hand it demonstrated the diversity of interest in the SO_2 market. Utilities purchased 99% of the Phase 2 permits for 2000 onwards – as could have been predicted from their general strategy of preparing for the rainy day when the tough Phase 2 reductions begin to bite. However, they were minority purchasers of Phase 1 permits. These were popular with thrill-seeking private investors (38%), brokerage firms (12%), public interest groups (9%) and 'others' (9%).

[26] Ibid.

The category 'public interest groups' refers mainly to environmental groups, purchasing permits which they 'retire' without their ever having been used. Some of the 'others' are undoubtedly entrepreneurs who are doing the same, but soliciting funds from the public through specialist environmental publications. In exchange for a sum of money, the customer gets a parchment certifying that a certain number of pounds of SO_2 has been retired.

As well as playing the market at the forced auction, the brokerage firms are filling a crucial role in providing information on the market price of permits. For example, the Emission Exchange of Denver publishes monthly price estimates. In January 1994, it set permits dated 1995 at $185 a tonne and those dated 2000 at $130 a tonne, at current prices. Brokers have also enabled trading to occur despite delays in the arrival of the EPA's permit tracking scheme, by arranging swaps between utilities wishing to exchange Phase 1 and Phase 2 permits, and also accepting conventional bids and offers for permits. One broker, Environmental Brokerage Services, has even encouraged producers of high sulphur coal to purchase permits to 'bundle' with the coal, thus making it more attractive to utilities.

Monitoring Efforts On the technical side, the first target of the TEP scheme was for those utilities covered by Phase 1 to install CEMs by the end of 1993. This was achieved by all except a handful of units, most of which were undergoing maintenance through the deadline period. Some utilities, unhappy with the performance of the CEMs, have gone beyond the EPA's monitoring requirements by installing secondary units. They are calculating that the increased accuracy and reliability of their monitoring will lead to lower certified emissions. They expect that the additional permits this releases for trading will offset their additional equipment costs.

Reactions of Public Utility Regulators One major unresolved concern still hangs over the TEP scheme. Because the US industry is based on private utilities operating local monopoly franchises, its electricity tariffs are micro-managed by state regulators, who pass judgment on how wisely a utility has conducted its business and determine the rate of return it can earn through the tariffs. There has been no clear guidance from the EPA on how the state regulators should treat purchases and sales of permits, and this uncertainty has dissuaded some utilities from trading. Utilities do not even know if they will be allowed to keep all the profit from selling surplus permits, or if state regulators will demand that some or all of this should be used to reduce electricity rates. The interactions between the utilities' rate-setting negotiations and the opportunities of the permit market may be so complex that the incentives for innovation become impossible to assess with any confidence. This issue is discussed further in the next section.

10.6.4. A Market for Innovation?

How successful is the US SO_2 trading programme likely to be as a means of stimulating innovation? Economic theories predict that the financial incentives for firms to innovate are very sensitive to the design of a trading scheme. Did the multitude of interest groups which were involved in formulating the trading programme have a clear sense of the impacts their decisions might have on innovation? How important is the structure of the utility industry and what conclusions can we draw from the behaviour of utilities to date?

Stacking the Deck The prospects for innovation were dealt a serious blow very early on. By choosing to allocate permits free of charge, rather than to auction them, the EPA adopted an approach which, according to some economists, encourages firms not to innovate in certain circumstances. These economists took a theoretical line which went beyond the earlier, simpler models of TEP schemes, on which the EPA and the Environmental Defense Fund based their proposals. The EPA did consider an auction-based programme, but decided that this would represent too much of an upheaval for the heavily regulated US electricity industry, since state regulators would face huge uncertainty about the amount and cost of generation available in their regions from one year to the next. Perhaps if the findings on the superiority of an auction system had been available earlier, the EPA might have explored solutions such as phasing in an auction system over a number of years.

The Dead Hand of the States The state electricity regulators represent another potential threat to the prospects for innovation. Consider the case of a utility faced with the choice of installing an innovative technology. If the innovation is successful, the utility expects either to have spare permits which it can sell, or to avoid having to purchase permits in the market.

In the first case, a successful innovation produces a stream of direct income from the sale of permits. Will the regulator allow the utility (i.e. its shareholders) to keep any of these profits? If the innovation fails, will the regulator view the investment as a speculative venture for financial gain, and prevent the utility recovering its investment costs from consumers?

In the second case, a successful innovation allows the utility to avoid purchasing permits. The utility regulator will almost certainly allow the utility to earn its standard rate of return on the capital cost. If the technology fails, the regulator may decide that the investment was unwise and refuse to include the investment in rate calculations, since a risk-free alternative (purchasing permits) was available.

In both cases, a utility's incentives to innovate are fundamentally determined by the attitude which the regulator will adopt once the success or failure of the innovation is clearly established. There are strong reasons for believing that regulators will err on the side of caution and that the interaction of the trading system with the tariff regulation system will create a disincentive to innovation.

One reason is that electricity tariffs are fundamentally a political issue. Voters are more inclined to punish their representatives for presiding over price rises than to reward them for price cuts. Electricity regulators, therefore, are under great pressure to avoid rate rises, but are less driven to seek rate cuts.

In addition, the very nature of permits reinforces this conservative approach. In the past, compliance was usually achieved by choosing from a set of technical options of varying risk. (Switching from coal to other fuels was often a political taboo, and is still outlawed by some state utility regulators.) By contrast, trading provides a guaranteed means of compliance at a reasonably predictable cost. This is likely to appeal to cautious utility regulators. Also, regulators do not need to understand technical issues when dealing with a trading system. There is a danger that this will promote a self-reinforcing trend where regulators initially seek to avoid risk by encouraging utilities to purchase permits. This might reduce their familiarity with technical issues, which in turn makes innovative technical options appear more risky and encourages further reliance on trading or standard technologies.

These considerations are not peculiar to the US system. They apply to any electricity supply regime where public officials have a powerful role, acting as an intermediary between the industry and the consumer. In a competitive industry the chances of a trading system producing excessive risk avoidance are less. Technical decisions would be made by those best able to do so, i.e. the generating companies, and there would be no political aspect to these decisions, only commercial ones.

Early Signs It is too early to draw any conclusion about the impacts of the permit programme on innovation, but some indicators are emerging. For example, the Ohio state regulator, the Public Utilities Commission of Ohio (PUCO), has strongly advised the Ohio utilities to engage in permit trading. The declared intent was to ensure that the utilities pursue least-cost strategies, but the strength of this message may reflect a bias within PUCO which favours purchase of permits over technical approaches.

The EPA has been on the alert for any signs of new willingness to pursue innovative approaches within the industry. So far it is not aware of any, apart from some efforts to maximize the performance of existing scrubber units. Some EPA officials are also pessimistic about the longer-term prospects for major innovations under the trading programme. Clean coal generation technologies such as coal gasification can deliver extremely low levels of SO_2 emissions, but have not yet been convincingly demonstrated. The demand for new baseload generation which is expected from 2000 onwards might also be expected to create a demand for clean coal technologies. However, it is just as likely that utilities will continue to extend the life of their power plants, by as much as 30 years. This, combined with the supply of permits banked up to the year 2000, could make clean coal technologies unattractive until 2010 at the earliest.

10.7. Conclusions

The main suppliers of emission control equipment for power stations are large engineering firms with significant political influence. This has created the conditions for a political linkage between emission regulations and national industrial policies. Unlike the study of vehicles in Chapter 9, where we were interested in the impacts of standards on a consumer product, this study introduces the prospect that policy-makers can use regulations to support environmental products and services directly.

Policy-makers are clearly responsible for creating the markets for NOx and SO_2 controls. This brings an additional responsibility to understand how firms will be affected by the markets they create. Combustion Engineering's experience with low-NOx burners in the United States illustrates how a firm's technological lead can be destroyed by regulatory uncertainty which interrupts the market and brings innovation to a halt. This is not an argument that environmental policy should be used to maintain markets when regulation is not otherwise justified, but that regulation should be used in a way which establishes as far as possible a stable predictable market, and hence appropriate conditions for innovation.

The advent of international regulation of acid rain has raised the spectre of countries attempting to create export markets for their pollution control industries. In the case of Germany's approach to acid rain in Europe there was clearly no such pre-existing 'first-mover' strategy. Germany had strong domestic reasons for its concern over acidification.

For policy-makers tempted to use regulations to maintain or create domestic markets and hence technological leads, some general objections arise from this study. First, this approach assumes too much about whether the same environmental concerns will manifest themselves elsewhere and how they will be addressed. Despite the existence of a framework (the EC) within which Germany exerted a strong influence over the timing and scale of reductions in other countries, this failed to translate into a significant commercial presence for German suppliers in the United Kingdom (the largest EC emitter after Germany made unilateral reductions) because the United Kingdom remained free to take a radically different approach to reducing emissions (fuel switching). Finally, large markets tend to have their own unique requirements and so provide the opportunity for domestic firms to develop the local, and sometimes global, state of the art. As the major effort to reduce acid emissions has shifted first from the United States to Japan and then to Europe, local firms have first adapted and then taken over the main control technologies.

Both of these features of the market argue against attempts to create strong industries through regulation, on prospects of exports. Far better to ensure that where regulation is required, it gives firms the opportunity and incentive to establish technological competence and a capacity for sustained innovation. Then if conditions abroad are favourable the domestic industries should, for a time at least, be strong contenders in those markets.

Through its SO_2 emission trading scheme, the United States has made an important advance in environmental policy. This is the first application of tradable permits on a scale large enough to provide conclusions which allow generalization about the operation and consequences of TEP schemes. Climate change policy, where many observers favour TEPs as a national or international instrument for reducing CO_2 emissions, should be one beneficiary.

Evidence so far from the US experience suggests that TEPs bring with them a new political dynamic and new institutional relationships, thus creating a market with its own peculiarities. In deciding on a TEP scheme and a national emissions cap early in the political process, policy-makers (including influential policy-oriented NGOs) shifted the agenda from the usual scientific and environmental debate to an argument over distribution. Political agents may have felt that since the overall implications of the proposed cap were not too threatening, they should concentrate on tactical manoeuvring within this limit. By contrast, the conventional approach in the United States often involves Congress making a legislative statement that some pollutant must be tackled, with the details to be settled later by the EPA and/or the states. Because the commitment is open-ended at the legislative stage it tends to invite outright opposition from those who are affected.

The TEP scheme has major implications for both the utilities and their state regulatory commissions. These long established institutions need to adapt their behaviour to the new situation created by the SO_2 market. There are signs that, in what they perceive to be their own political interests, the state commissions may undermine some of the TEP scheme's incentives for utilities to innovate.

Part 4

Chapter 11

Environmental Policy, Industrial Strategy and Innovation

11.1. Introduction

The country studies comprising Chapters 3 to 8 illustrate the great diversity of attitudes and approaches to setting environmental objectives and dealing with industry. However, we can also draw useful general lessons from this material. This chapter presents a transnational analysis of the political climate of environmental policy and the processes of dialogue involved in environmental regulation. The purpose of this analysis is to provide an insight into why environmental regulations have been a source of conflict in some countries and not in others; why they are sometimes associated with costs and burdens, and at other times (in other places) with technical progress and innovation. Key themes which emerge are the stability of environmental policy and the nature and extent of dialogue between industry and regulator.

Short analyses of the vehicles and clean coal case studies allow us to examine these issues from industry's viewpoint and to draw general lessons for policy-makers and industrialists alike. Policy instability emerges as an important element in industry's perception of commercial risk. An unstable policy climate causes distrust and pushes industry towards misusing dialogue mechanisms in an attempt to manipulate or mislead regulators.

In the vehicles industry, lean production has been responsible for introducing an entirely new logic to volume manufacturing processes. This is transforming firms' internal capacities for innovation. Policy-makers need to learn to recognize such transformations, and work with them. Technologies for cleaner coal fired power generation have been driven hard by regulations, but display a remarkable fluidity and adaptability to local conditions. Most of all, this example illustrates that the nature of the technologies which a market will demand is unpredictable, and cautions against hopes of capturing large foreign markets through a 'first-mover' strategy.

11.2. The Political Climate of the State–Industry Relationship

Where environmental issues are concerned, political actors are caught between a rock and a hard place. On the one side is industry, concerned about costs and constraints on its activities, on the other is the private individual or citizens' group. How does environmental policy reflect this tension? The evidence from the national experiences in this study suggests that there is no simple relationship between the strength of environmental feeling and the effectiveness of the national system of environmental protection.

Neither is there any simple link with the political power of industries. In some countries, environmental issues follow the pattern of general state–industry relationships. In others, this relationship has a distinctive character, where environmental issues are concerned. Any attempt to describe the political climate for environmental policy by applying accepted wisdom about the relationship between state and industry is likely to be misleading. For example, in Japan the national government is supportive and even protective of manufacturing industry, in particular. However, the national climate of environmental policy is largely determined by the actions of powerful local authorities which put a high priority on protecting their citizens.

Over several decades, these differences in political attitudes have evolved into distinctive political climates, even in countries which experienced similar political turning points, such as the 1968 peace protests. The political climate of environmental policy has a profound effect on the attitudes and behaviour of firms which are subject to environmental pressures. This in turn determines the nature of regulatory regimes and the range and flexibility of instruments available to regulators, and these detailed aspects of policy implementation can act to reinforce or undermine the political climate.

This section draws out the distinctions between political climates, and examines the forces responsible for their evolution. The political independence of the state from industry – *where environmental issues are concerned* – is perhaps the most important factor differentiating the countries in this study.

11.2.1. France: L'état, c'est l'industrie

First, France, at one extreme, has highly centralized political processes. Not least of these is the five-year plan for the economy, which focuses on those industries which are publicly owned, such as car and computer manufacturing, airlines, energy and banking. Although things are now beginning to change, partly as a result of European Union pressure for liberalization, for many sectors of industry it remains the case that the state *is* industry, and vice versa. Even in those areas where the ties between the two have been loosened, officials retain their sense of ownership. French politicians of all hues share a sense of the strategic importance of certain industries which is regarded with envy by politicians with socialist leanings in many other countries. French industrial prowess is a key element of the national sense of possessing an exceptional culture (*exceptionalisme*), a belief which remains strong and is explicitly referred to in the popular press even today.

In France, therefore, industry must be seen as being a part of the state and at the heart of the state's concerns, to a greater extent than in any of the other five countries examined. Environmental concerns have been slow to find official expression. Where they have, it has often been at the local level, and solutions have been found which entail no repercussions for national authorities. In sum, the centralization of power and authority in Paris and its intimate association with industry has defended industry against all threats, just one of which is concern for the environment.

11.2.2. Japan: Citizen Power

For the Japanese, industry is in many respects a national champion just as it is for the French. But the history of the state–industry relationship in Japan is a complex one which has suffered many reversals, and historical events have prevented industry from becoming intimately associated with political authority, despite the contrary view of many western observers.

Japan went through a remarkably swift industrialization in order to catch up with the West, following the Meiji Restoration in the mid-nineteenth century. This industrialization was led by family dynasties. After the Second World War, the freedom of firms was curtailed by legislation. Comprehensive workers' rights made dismissal impractical for most large firms. In response, the philosophy of lifetime employment was created and strict labour relations, rather than being a source of conflict, were transformed into an advantage. This in turn determined the nature of the state–industry relationship in modern Japan. Instead of drifting back towards the political realm of the state, the corporation became the community, and the community, the corporation.

Another postwar legacy was a relatively weak and ineffective parliamentary structure. Authority rested instead with the public servants. As with most democracies where the civil service is apolitical, public service attracted those with a will to make a better society. In this they found common ground with a community-oriented industrial sector. Preserving communities and the well-being of society in general became a guiding principle for policy-makers, and one which industry saw as being, at the very least, a sensible objective. Where there is a need to resolve a conflict between environmental concerns and industry, Japanese officials are guided by the principle of what is best for the community they serve.

During the 1960s, the large metropolitan authorities took the lead in tackling severe pollution problems. Shocked by environmental scandals, their main concern was for the health of their citizens, at a time when national authorities were more concerned with patterns of growth and industrialization. Community-oriented firms cooperated with the metropolitan authorities and new regulatory mechanisms evolved. These included formal networks for communicating best practice (both technologies and regulatory approaches) between these authorities. As a consequence, the national climate for environmental policy came to be based upon a high degree of political independence from the polluting industries. In the late 1960s and 1970s, national environmental legislation tended to follow the path mapped out by the metropolitan authorities.

Inevitably, this high degree of independence from industry on environmental policy is prey to political tensions. However, the mechanisms adopted for the detailed implementation of policy (about which more will be said later in this chapter) avoid and reduce these tensions through their emphasis on close involvement of industry. This lessens the incentive for industry to make political or legal challenges to environmental policy and, over time, the lack of such challenges builds up the credibility of environmental policy and further enhances its independence from industrial interests.

11.2.3. The United States: The Hippie Schism

In the United States, rule-making and legislation have, until recently, been everything. An obsession with defining the state–industry relationship in excessive detail, arising from a lack of trust, has led to tedious and stifling rules for every stage of the legislative process. Pursuit of perfection has truly been the enemy of the good, as the inability to communicate freely has intensified and prolonged the schism between industry and regulator which followed the turbulence of the peace movement and hippie culture of the late 1960s.

The 'hippie schism' and the country's response to it is the defining feature of the state–industry relationship in the United States. In a short time, industry went from being the bringer of untold riches, to being a despoiler of nature, a poisoner and an ally of the devil. A whole generation turned against the values of its parents and the industrialists retreated to their bunkers in shock. With no reconciliation in sight, they soon came out fighting.

Politicians openly used the environment as a 'wedge issue', with which to define their difference from the opposition. Rules excluding those with first-hand experience of polluting industries from being employed by certain regulatory agencies, legislative hammer clauses and legislative prescription of the institutional structure of regulatory agencies, ensured that there could be no meeting of minds between regulators and industry. An adversarial mindset was institutionalized.

This has had the effect of establishing a relatively high degree of independence of the state from industry. Unfortunately, it has also helped to establish environmental policy as a political battleground. During the early 1980s, Ronald Reagan's deregulation policies and his appointments to the EPA weakened this independence, although a Democratic Congress held this in check. Reagan justified his policy largely by reference to regulatory absurdities which were a direct consequence of the prescriptive and adversarial relationships previously established by Congress.

11.2.4. Germany: Uniqueness of the Environmental Issue

Within Europe, America's hippie schism reverberated most strongly in West Germany. The party of power throughout the 1970s first embraced environmental concerns but soon reverted to its roots as the representative of industrial workers. Environmental issues became associated with local activism and the national peace and anti-nuclear movements, while the national environmental institution, UBA, remained weak and remote from political power. Forest death was taken as dramatic verification of the doomsday scenarios of the environmentalists, casting industry as the enemy. Radical environmentalists acquired political power, through the creation of an Environment Ministry, and consciously distanced themselves from industry.

This is an aberration in the general state–industry relationship in Germany. For the most part industry is well represented in the political process, the nature and needs of industry and commerce are well understood by politicians and policy-makers; and the Ministry of Economic Affairs holds a position of substantial power and authority. But the environment is different. Many German policy-makers seem to view an adversarial relationship between environmental policy-makers and industry as a natural state of affairs. Germany's most influential Environment Minister, Klaus Töpfer, was generally applauded for his hard-line, anti-industry rhetoric. Only the recession of the early 1990s has caused some to re-think the wisdom of this approach.

Germany has, like Japan, established a strong political independence between state and industry on environmental issues. However, in achieving this it has limited the options available to regulators, an issue we will return to shortly.

11.2.5. Denmark: Luck or History?

Denmark made conscious efforts to avoid the 'hippie schism' rolling through North American and European politics. Neighbouring Germany was setting a powerful example of confrontation, but, partly through the vision of one person – the first Minister for the Environment – and partly drawing on a tradition of consensual politics, industry was deliberately integrated into environmental policy formulation. Over time, this combination of luck and cultural predisposition allowed trust to grow and schism was avoided.

Placing a high priority on environmental protection has become an article of faith shared equally by all political parties in Denmark. As a result, environmental protection has gradually taken on the character of a fundamental social policy, similar perhaps to the commitment to a minimum level of welfare for all citizens. This has evolved to the point where the Environment Ministry is second only to the Finance Ministry in its political power. The most recent demonstration of the political importance of the environment was in October 1994, when responsibility for energy policy was transferred to the Environment Minister. This has implications beyond

Denmark: in the Council of Ministers of the European Union there is now one voice arguing from the environmentalist's point of view.

Once again, the specific mechanisms for implementing environmental policy emphasize cooperation and consultation with industry. As is the case in Japan, this tends to increase the credibility and independence of environmental policy over time. In Denmark, these mechanisms rely in large measure on the personal relationships between the relatively small groups of individuals having expertise in any one industry/environment area. These individuals tend to have mobile careers in which they may move frequently between government, industry and academia.

11.2.6. The Netherlands: Towards a Planned Environment

In the Netherlands environmental politics initially followed much the same path as in the United States, Germany and many other countries. Industry tackled the worst sources of pollution, but this was seen as an unwelcome imposition. Within the government, familiar battle lines were drawn between industry and environment ministries, and environmental policy was by no means free of political favouritism towards industry.

In the mid-1980s, the government initiated a long-term programme consciously to change this relationship and actively to involve industry in creating the solutions to environmental problems. A dispassionate inventory of the extent of the continuing environmental damage to national ecosystems (despite a massive clean-up of the gross and obvious polluters over the previous decade) had created a political will to set firm, long-term national targets for pollution reduction. The explicit aim was to transform the Netherlands into an environmentally sustainable economy.

One of the most striking features of the resulting National Environmental Policy Plan is that it has been consciously designed to create forms of relationship between industry and state characteristic of those which evolved in Japan and Denmark through a combination of chance and cultural predisposition. The long planning time scale and the use of ambitious national targets which have been negotiated with industry are designed to ensure that environmental policy is stable and consistent over the long term. This policy stability should in turn ensure the state's political independence from industry on environmental issues.

11.2.7. Some Lessons: Policy Stability, Trust and Innovation

A political climate where environmental policy is stable over the long term and is formulated independently of special pleading by industry reduces the range of options available to industry. With influence-peddling in all its forms effectively ruled out, industry is more inclined to focus on the technical nature of the problem and to devote its resources to innovation. With policy stability, a sense of trust in the motives of politicians and officials slowly develops. Firms begin to feel willing to disclose the potential for technical improvements to solve environmental problems, and legislators can incorporate this information when they formulate regulatory instruments. This model of the state–industry relationship avoids the trap of power-based negotiations – where industry seeks to use its superior information to its own advantage – and evolves instead into a relationship based on open exchange and information sharing.

Comparison of the six countries in this study suggests that creating this stable, trusting political climate is the first stage in harnessing innovation to meeting environmental objectives. This process is completely different in character from 'technology forcing', the catch-all phrase which is often used to suggest that some legislators create standards which will force industry to innovate and create new technologies. References to technology-forcing generally assume that legislators have a blind faith that science will inevitably provide the answers. In practice, legislators in the countries studied here never adopted this approach, with a single exception – electric vehicles in California – where it appears that the state regulators were misled by, or perhaps misinterpreted, the actions of one auto manufacturer.

Policy stability and trust are beneficial in other ways. The political and regulatory climate is an important part of the business environment within which firms operate. The role of the state–industry relationship in influencing corporate strategy, specifically the decision to pursue innovation, is discussed later in this chapter. The next section addresses the detailed functioning of regulatory regimes, where the background state–industry relationship has a major influence on the extent to which innovation is allowed to play a role.

11.3. Processes of Dialogue in Regulatory Regimes

Superficially, the day-to-day implementation of environmental regulations in each of the six countries studied has a great deal in common. Every country has some form of industrial site licensing system, constituting the main element of the regulatory regime for local pollutants. A body of officials, either belonging to a national inspectorate or employed by local authorities, is responsible for setting specific standards for industry, usually following broad objectives handed down by the legislature. In most cases, the same body of officials ensures that industrial firms comply with existing regulations on types of equipment, handling of pollutants, absolute levels of emissions or concentrations of pollutants in waste streams. But these apparent similarities in regulatory regimes conceal differences which are evident in the details of setting standards and implementing them. These differences in detail are themselves symptoms of fundamental differences in philosophy which are intimately connected to the political climate of environmental issues and which determine the extent to which innovation provides a solution to environmental objectives.

This section looks at some of the common features of the regulatory regimes in the six countries studied (organized by subject, rather than by country). The unifying theme here is the quality of the dialogue between industry and the regulator: the nature and extent of dialogue, the competence of officials, the willingness of firms to part with information. Through engaging in a process of high-quality dialogue, certain regulatory regimes have achieved a sensitivity towards industry which allows innovative solutions to emerge, and created robust relationships which can survive and even defuse tensions in the political relationship between the state and industry.

11.3.1. Political Constraints on Processes

Standard-setting processes vary from the formulaic, procedure-driven approach in the United States to a flexible, personal approach in Japan or Denmark. Strict rules govern the ways in which officials of the US Environmental Protection Agency interact with industry. Regulations are often drafted following little or no consultation with outside groups, prior to publication for comments. At the post-publication stage, environmental lobby groups represent any changes proposed by industry, for whatever reason, as a climb-down by the EPA.

These procedures are imposed by Congress. They perpetuate the adversarial political relationships surrounding the legislative process in Congress itself. Frequently, the result is inappropriate, costly standards which create bad feeling in industry, reducing its willingness to cooperate. In a classic vicious circle, this lack of cooperation leads to more inappropriate regulations.

In Japan, by contrast, the primary regulator responsible for determining environmental standards and therefore shaping the regulatory regime is the metropolitan authority. Regulators have intimate knowledge of the industries they deal with and close relationships with industrialists. Because of the climate of policy stability and trust, they can pursue flexible regulatory approaches which would be open to abuse in other countries.

11.3.2. Why Japanese 'Technical Hearings' Are Not a Liar's Charter

One of the most distinctive features of the Japanese standard-setting process is its reliance on a multilateral flow of information between small and large firms, national and local government agencies, academia, etc. The system of technical hearings, where firms discuss, with regulators, current and potential technical progress towards better environmental performance, is the most visible manifestation of this process. Of course, similar discussions occur in most other countries, but generally these are on an ad-hoc basis and are open to the possibility that the firms will intentionally mislead the regulators. Technical hearings are formal requirements held at regular intervals, often annually, and there does not appear to be any significant problem with the quality or trustworthiness of the information which is volunteered.

One Japanese executive from a multinational manufacturer offered the explanation that it would be unthinkable for him to mislead or withhold information from Environment Ministry officials. Although there are elements of a fundamental respect for authority in this attitude, at least part of the explanation lies in the fact that the Japanese firm is attuned to the wishes of the community (which it feels itself to be a part of) and agrees with the social objective of environmental protection.

Just as importantly, firms have learned over time not to fear the uses to which the authorities will put the technical information they provide. This outcome rests on the competence of the officials, who are sufficiently knowledgeable to avoid using the information to devise regulations which would impose unnecessary cost or restrictions on industry. Finally, at the local level especially, firms recognize that regulators see protecting the community as their primary purpose, and have been consistent in this over many years. In these political circumstances, special pleading is deemed to be ineffective, so reinforcing a corporate mindset which associates the firm with the community, and encouraging cooperation with the technical hearing system.

11.3.3. Flexibility and Iterative Processes

The regulatory path followed by the Japanese local authorities, based on the information supplied by industry, is flexible, allowing scope for retreat if technical progress does not live up to expectations. An initial period, where the putative guideline is merely a 'recommendation', will typically last a few years. This period will be in keeping with natural rates of turnover of capital stock or product lifetimes, where these are likely to be affected by the new standards, and developments by third party equipment suppliers. The continuing technical hearings, and other frequent consultation, allows the results of firms' early efforts to meet the standards to be taken into account, when the final mandatory regulation is adopted.

Through experience, firms know that the mandatory regulation will be adopted at the expected time, so there is little incentive to do nothing. Instead, firms experiment with a variety of ways of meeting the standards, often through a series of small continuous improvements and 'good housekeeping' measures. This process is characteristic of Japanese firms, many of which strive for continuous improvement, or *kaizen*, in all aspects of their production.

Competence, therefore, is essential on both sides if such a flexible, iterative approach to regulation is to succeed in allowing innovation to play a major role in meeting objectives. The flow of information in itself is not sufficient. In Japan, the regulators are technically competent to seek appropriate information from industry and to use that information to devise appropriate putative targets and regulations. Japanese firms are already competent in their management of the process of innovation, and capable of applying that expertise to environmental improvements, when a regulator sets an appropriate challenge.

11.3.4. Regulate Locally, Innovate Locally

Japan's experience also illustrates the benefits of local responsibility for formulation of standards and regulations, compared with national control. Local agencies are better able to take account of the unique circumstances of each site, such as the effects of pollution, the technical capability of the firm and its financial health. New technologies can be developed and demonstrated where the need is most pressing. Conscientious efforts to exchange information between local authorities ensure that the best solutions can be adopted elsewhere, once they have been proven.

Advances which are generated locally tend to filter upwards until they are incorporated into national legislation. In other countries, the reverse is more often the case, with national requirements imposed where they may not be appropriate. Long delays in national legislatures and regulatory bodies frequently mean that measures which once made sense are inappropriate when they are finally enforced.

11.3.5. Compliance vs. Cooperation

In common with any other country, there are those Japanese firms which are reluctant to comply, even after recommendations become mandatory. These are almost always small firms. As a result of close contact throughout the formulation and trial periods, regulators can discriminate between those firms which have genuine difficulties – where further advice and even financial assistance may be offered – and those which are simply recalcitrant. For most industries, however, the working assumption is that the great majority of firms will cooperate willingly. Elaborate systems of penalties are therefore unnecessary.

In the United States, by contrast, history has shown that industries as a whole will fight legislative proposals, oppose and frustrate regulatory processes and challenge the final outcomes in court. Many EPA regulators assume that most firms will do everything to avoid complying with regulations at their industrial sites. In fact, there is a difference of perception here. Local EPA officials, or officials of state regulatory bodies, report that firms are generally willing to cooperate and comply, but are frustrated by the nature of the regulations and their lack of influence during formulation of these. Until recently, this willingness to cooperate has not been effectively communicated to national EPA regulators, who have focused much of their energies on mechanisms for forcing compliance, nor to Congress, which has persisted with the restrictive legislation to which firms object.

Elsewhere, some but not all of the useful features of the Japanese approach are apparent. In Denmark, the community is involved from the outset in formulating responses to environmental issues. During early consultation, political parties, officially recognized environmental and citizens' groups, academics and industry all discuss the best legislative approach. Although regulations tend to be generated at national level, the appropriate comparison is with Tokyo, which has a larger population than Denmark, rather than with the United States. Final standard-setting is dependent on knowledge and experience arising from the joint innovation efforts of government, academia and industry. Individuals often shift between these realms, promoting a common understanding and creating a basis for trust.

There is not the same trial period for firms to find innovative responses, although more formal government-assisted demonstration programmes mimic this feature of the Japanese approach to some extent. There is also less feedback from the Danish inspectorate to policy-makers. By contrast, this happens automatically in Japan, where local authority officials fill both roles. However, the Danish system of allowing officials from the local pollution inspectorates to offer free consultation and advice on how firms can exceed existing requirements and anticipate new ones, helps to spread best practice. A recent Finance Ministry proposal to charge for this advice threatens this mechanism. This is a good illustration of the need for the state to deal with industry through competent, knowledgeable officials and of how failure to do so can lead to a breach of hard-won trust.

Summarizing what we have learned about those elements of a regulatory system which allow industry's capacity for innovation to be effectively harnessed, one conclusion stands out clearly: a high-quality flow of information between industry and the regulator is essential. For this dialogue to exist, regulators must be technically competent and trust must be established; otherwise firms could attempt to mislead the regulator. There must also be some sense of inevitability that regulations will be enforced; otherwise firms will not be encouraged to seek innovative solutions and some will try to evade or derail the regulatory process. Once again the key to this is long-term policy stability and independence of the regulator from special pleading.

11.4. Lessons from the Industrial Case Studies: Market Needs and the Capacity for Innovation

Both of the industrial case studies in this book were chosen for their political importance as key industries which Europe, the United States and Japan have fought over, amid accusations and counter-accusations of attempts to manipulate environmental standards for commercial ends.

In each of these cases, certain governments and firms sought to make a direct link between environmental regulations and the international competitiveness of domestic industries. Close examination of the effects of environmental regulations on the industries concerned suggests that there is no simple link with competitiveness. These examples serve as warnings against trying to create competitive advantage by manipulating national or international standards. Instead, they emphasize the importance of being clear about environmental objectives and then devising a regulatory approach which allows industry to harness innovation to meeting the desired objectives.

11.4.1. Vehicles

The car industry is the birthplace of the mass production system of manufacturing. It has traditionally been dominated by enormous multinational firms reaping economies of scale. Its products are marketed to consumers on a combination of technical performance, reliability, styling and brand name. Chapter 9 traces the battle over standards for vehicle exhaust emissions, from the first serious attempts to tackle auto pollution in the early 1970s to the European regulatory process culminating in the early 1990s. It is the story of how a revolution which is transforming the modern industrial world – the advent of lean production – has generated paranoia in Europe and the United States, with a host of misinterpretations and misunderstandings, only one of which is the strange notion that emission standards might in themselves dictate the success of major car producers.

Mass Production – Mass Delusion In the 1960s, the Japanese car industry was organizing around the new manufacturing paradigm of lean production while the rest of the developed world was mired in the inefficiencies and rigidities of mass production. In the American and European firms, set-up costs were enormous, production runs had to be vast to recover these costs, change was feared and innovation was studiously avoided. When the Japanese adopted tougher emission standards than those in the United States, American producers convinced themselves that this was a ploy to force enormous costs on them – costs which Japanese firms were mysteriously in a better position to absorb. They convinced their politicians too, and the United States publicly attacked this 'unfair' trade practice.

Internal Strength Despite special treatment for dirtier American vehicles in the late 1970s, US auto manufacturers continued to ignore the Japanese market and made only a pitifully small number of sales there each year. The Japanese firms knuckled down to the task of adding three-way catalysts to all their models. Lean production techniques and a philosophy of innovation through continuous improvement helped to make this a relatively painless task. More importantly, this attitude to innovation led to an increasing gap between the quality and performance of Japanese cars and their American rivals. The Japanese encroachment on the US market continued, with the issue of emission standards having no discernible impact one way or the other.

In their dealings with their own regulators, the Japanese firms had a confidence in their own innovative abilities. The larger, long-established firms which were initially sceptical did not try to pull any strings or fight higher standards, once their smaller rivals had demonstrated the technical feasibility of three-way catalysts. Within a year or two their competence in managing innovation delivered commercial implementations of the new catalytic converters.

Europe: 1980–90 In Germany, commercial considerations persuaded the auto manufacturers to press for a European strategy based on catalytic converters, rather than better engine design. In part this reflected their adherence to mass production: economies of scale would be greater for firms such as Volkswagen if they could provide identical models for the European and American markets.

So began a long and bitterly politicized battle over European emission standards, a battle which ended in irrelevance. By the early 1990s, the European and American auto industries had begun to realize that the Japanese had killed off mass production as an organizing principle for manufacturing industry. In the context of vehicle emissions, having a head start in catalytic converters, on the one hand, or lean-burn engines, on the other, was no longer important. Being a flexible, innovative manufacturer was the key to success. Rover, the British manufacturer which was expected to be the big loser in this standards battle, formed an alliance with the Japanese, adopted the new lean production manufacturing techniques, and led the European car industry out of recession.

Electric Vehicles The extraordinary story of the genesis of California's Zero Emission Vehicle (ZEV) legislation, which will have the effect of requiring electric vehicles to be offered for sale from 1998, exposes a great falsehood underpinning the popular notion of 'technology-forcing' regulations, i.e. the view that environmental standards should be manipulated in order to force new technological discoveries. Technology-forcing has not in the past been a favoured strategy for any policy-maker, even those in California. It has become the justification for the ZEV legislation only as an attempt to cover up the misunderstandings and confusion which surrounded the policy-makers' initial decision.

From the perspective of corporate strategies towards technology and environmental standard-setting, the ZEV story serves as a warning of the dangers inherent in adopting a tough negotiating stance with policy-makers. A long history of mistrustful relationships between the big auto manufacturers and the Californian regulators created an atmosphere which allowed confusion to flourish. It matters little whether GM was hoping for a ZEV regulation, as part of a grand first-mover strategy, or whether it was merely showing off its achievements with the *Impact* and wished to be left to market it at its own pace. The regulators' relationship with the big producers was too weak and lacking in trust for the truth to be established, from either GM or its rivals.

11.4.2. Clean Coal

The market for emission controls for coal fired power plants is dominated by large, politically important heavy engineering firms selling to well informed customers in the electric utilities. Utilities, often publicly owned, have tended to favour domestic suppliers. The emission control equipment often has to be adapted to domestic market requirements and is frequently tailored to each implementation at each plant.

As a result of this combination of features, innovation has followed closely the needs of the market and the technological lead has been passed from one region to another, as the major markets have shifted from region to region. FGD was first attempted in the United Kingdom, as a possible response to London's smogs, but was not in the end required and was not pursued. First the United States, and then Japan, commercialized FGD processes when they needed to tackle their urban air quality problems; then Germany took the lead as part of its massive effort to control acid rain. Similarly, Combustion Engineering in the United States all but collapsed when the entire US power generation market foundered in the mid-1980s. International Combustion in the UK licensed and then modified Combustion Engineering's low-NOx burner, to meet the demands of the UK market. International Combustion grabbed a large part of the US market as activity picked up in the wake of the 1990 Clean Air Act, but many US rivals are now offering LNBs, and IC's market share will almost certainly stabilize to a less dominant level.

The emission control equipment market has been heavily influenced by the lack of free international outlets. Nevertheless, the variety and complexity of the technologies and the variability between countries suggest that the pattern of large national markets generating indigenous technology suppliers will continue.

This may be bad news for some existing technology suppliers in Japan and Germany and, increasingly as the 1990 Clean Air Act requirements take effect, in the United States. Firms which hoped that their technological leads (established on the back of large, domestic regulatory-driven markets) would translate into world domination of markets in industrializing countries, are likely to be disappointed. Ease of knowledge transfer and variability of local market conditions are likely to ensure distinctive local solutions, wherever a substantial domestic market exists. Firms such as ABB, which cultivates local subsidiaries with distinctive national identities and their own research and development potential, are likely to enjoy the greatest success.

One of the starkest lessons of the clean coal story is the folly of trying to predict market opportunities, and then basing regulatory decisions on the prospects for future business. The record of the genesis of the German acid rain regulations, and subsequent European regulations, shows that German policy-makers were motivated almost exclusively by environmental concerns. However, if, as some people believe, they set out to create markets for German pollution control equipment, particularly in the United Kingdom, then they failed. Other countries will generally hold many different policy levers which can be used to meet international obligations. In the United Kingdom, privatization of the electricity supply industry paved the way for fuel-switching on an unprecedented scale. In other circumstances, different options are likely to be available.

Recently, American regulators have introduced a comprehensive tradable emission permit (TEP) scheme, as part of a national cap on SO_2 emissions. Some of the early lessons for regulatory design and innovation are discussed in the next chapter. In terms of applicability, it seems likely that TEPs are best suited to industries with a relatively small number of large firms, capable of devoting substantial management effort to the financial complexities of futures, options, swaps and other trading paraphernalia. Even with their vast resources, some US utilities have initially shown a lack of confidence in participating in the new market.

11.5. Summary

How does an individual firm or an industry, acting collectively through trade associations or other lobby groups, respond to a new environmental issue or to indications that regulators intend to pursue new environmental objectives. Will they obstruct, ignore or cooperate? Will innovation play a part in their response? This study has found very different answers in Denmark, France, Germany, Japan, the Netherlands and the United States. However, in each country the distinctive nature of the political relationship between industry and government on environmental issues, and of the processes of dialogue between industry and regulator play a large part in determining the role for innovation.

For policy-makers it is important to know whether the successful features seen in one country could be applied elsewhere. The answer depends very much on one's opinion of the importance of culture in determining what policies can be successfully implemented. It is certainly true that both Japan and Denmark are often said to be culturally predisposed towards cooperative relationships in commerce and politics. Nonetheless, mechanisms such as the Japanese technical hearings are essential for maintaining industry's close involvement in regulatory processes. Favourable cultural conditions may have provided the breathing space for a flexible regulatory approach, based on a high-quality industry–regulator dialogue, to develop initially. This does not mean that these techniques cannot be applied elsewhere.

Inevitably, these cooperative processes would at some point come under pressure, if they were implemented without an accompanying political commitment to the environment, such as has occurred in the Netherlands. But the cooperative approach itself eases this political shift. However, for those countries which have traditionally relied upon adversarial processes, industry is unlikely to adopt new ways of dealing with regulators, without a large dose of political vision and leadership.

How do firms perceive the actions of policy-makers and what are the implications for innovation? The general climate of environmental policy is at least as important to a firm's willingness to innovate as the details of regulatory processes. When environmental policy is unstable, firms perceive the business environment to be volatile and therefore risky. Innovation is an inherently uncertain activity. It is subject to doubts about, for example, final technical performance or unforeseen market barriers, which are not present in other commercial activities. It is coming to be accepted wisdom that firms feel much more comfortable pursuing innovation when the general business environment, such as macro-economic policy, is stable. For industries where the environment has a significant impact, creating a stable environmental policy over the long term plays a major part in reducing risks and uncertainty and so encouraging innovation.

In the final chapter, following the next, a simple model will be used to integrate the two fundamental factors which have been explored above, i.e. the political and the procedural aspects of environmental policy. This model illuminates the relationship of these factors to one another and allows us to explore a range of possible strategic policy options.

Chapter 12

Policy Instruments

12.1. Introduction

This chapter puts the previous discussion into the context of current mainstream debates about environmental policy. In particular, it focuses on the continuing political debate over the effects of different types of regulatory instrument on industry, especially their effects on innovation. The other major theme addressed here is 'partnership' between industry and the community, and the role of socially responsible firms. Voluntary agreements are seen as a promising form of partnership for dealing with environmental issues. Several examples of voluntary agreements have been encountered in the course of this study. They throw light on the conditions for successful agreements, their relationships to other regulatory instruments and their effects on innovation.

12.2. Conventional Regulatory Approaches

12.2.1. Emission Standards

Many environmental regulations are expressed as an emission standard of one form or another. Generally, the acceptable level of emissions is related to the basic polluting activity, for example, grams of NOx emitted per kilometre travelled by a car, or concentration of NOx in the flue gases in boilers of certain sizes, using specified fuels.

However, emission standards do not by themselves control the total output of pollutants, unless the scale of the polluting activity is also limited in some way. Emission standards are, therefore, frequently augmented by total emission control regulations, which allow regulators to set and enforce absolute limits on total amounts of pollutant being discharged, into the atmosphere in a certain region, into a given watercourse, or lake, and so on. This approach is simplest where the polluting activity is dominated by a homogeneous, highly concentrated industry, such as electricity generation; regulators need only deal with a small number of actors.

Instances where an absolute environmental objective has been established, in terms of the total pollutant load which the environment can accommodate, are the exception rather than the rule. Most regulations have been formulated on the basis of uncertain information about effects. The regulator's task is to reduce emissions as much as possible, with minimum economic impact. Given these uncertain environmental objectives, regulators have tended to respond in two ways. At one extreme, they choose a formal emission standard, and a deadline, and then force industry to comply. At the other, the regulator sets a goal, encourages industry to make 'best efforts', observes the results, and then draws up a series of formal emission standards as 'back-markers' for the progress already achieved.

In both of these cases, the emission standards, as codified in legislation or local authority rules, may appear identical. However, their roles in the process of controlling emissions are very different. In one case, the formal regulation initiates the process, and will therefore tend to constrain all subsequent activities. In the other case, the formal regulation is the end of the process, or a step on the way, acknowledging the progress to date.

So, do emission standards stifle innovation? In the case of standards which initiate a process the answer is: yes, if circumstances change and regulators cannot quickly and easily modify the standards in response. In contrast, standards which acknowledge progress do so regardless of whether innovation was a significant factor in achieving that progress. However, they play a role in convincing industry that regulators will force laggards to act and therefore, over time, they are an essential part of establishing a climate where firms (polluters and third party equipment manufacturers) naturally consider innovation as part of their response to the overall goal set by the regulator.

12.2.2. Performance Standards

Performance standards are regulations which specify that a pollutant must be tackled by equipment which acts in a certain manner, reducing the pollution by a specified amount. In some cases, regulations (or their interpretation by officials) may name a specific piece of equipment as being the only acceptable method of treatment. It is this kind of regulation which is often referred to as the 'command and control' approach, although this phrase is also applied to emission controls.

In practice, many performance standards have evolved into *de facto* requirements for particular items of equipment. Local officials learn, for example, that a certain model of digester, supplied by firm X, reduces the pollutant in an effluent stream flowing at a particular rate to the required concentration. These officials will often lack the resources to evaluate new solutions, if larger national or regional agencies fail to do so. They will only grant operating licences or inspection approvals if the company buys the tried and tested equipment. The technology becomes locked in, stifling innovation. This same process has also caused emission standards to evolve into equipment requirements. Regulatory regimes which set standards up front, and then fail to update them regularly are prone to this. One example is the 'TA Luft' air quality regulation in Germany, which has remained unchanged for periods of up to ten years. Chapter 5 showed how legislative aspirations such as the TA Luft's 'state-of-the-art' clause can be so frustrated by conservative regulators as to render them almost meaningless.

As with emission standards, the role of the standards in the overall regulatory process – as initiator or as 'back-marker' – determines whether innovation is likely to suffer. However, the risk of stifling innovation is greater with performance standards than with emission standards, largely because it is more difficult constantly to update large sets of technical specifications than to revise emission standards. Many regulators have, therefore, tried to avoid performance standards in recent years, recognizing that they characterize the inflexible, and therefore costly, meddling most resented by industry. Nonetheless, one section of the 1990 US Clean Air Act which is guaranteed to induce nostalgia for the old ways includes detailed specifications of the percentages of SO_2 emissions which the seals and doors of coke ovens can each contribute to the overall emissions from the ovens.

12.3. Economic Instruments

12.3.1. Environmental Taxes

Industry expresses some serious concerns regarding the effect of taxes on innovation, arguing that innovation frequently requires capital for investment.

Pollution taxes siphon working capital from firms and make new investment more difficulty and/or more costly, if it becomes necessary to borrow the capital from external sources such as banks. Recycling tax revenues as investment subsidies for less-polluting equipment and processes can be useful in overcoming this disadvantage. However, most western finance ministries are wary of such schemes, on the grounds that it is difficult or time consuming to establish that the proposed investment is truly additional and would not have occurred even without the subsidy.

As with other kinds of regulation, a fundamental difficulty occurs in timing, and in allowing firms to anticipate new goals and react accordingly. Announcing a tax in advance allows industry to plan investments and innovation, and avoid the direct financial consequences of the tax. (This is, of course, distinctly unhelpful to those parts of government which see eco-taxes as revenue-raising, rather than environmental, mechanisms.)

Determining the precise effect of a tax presents special problems. First, there is the difficulty in establishing the tax rate which will cause the pollution to fall to the desired level. Initially, a tax may cause a sharp reduction in pollution, as firms take no-cost action, such as better 'housekeeping' measures. Later reductions will depend on the available set of pollution reduction technologies, which may be added to over time, and on alternative production processes. So taxes are blunt instruments even where we are only considering the primary objective of achieving some environmental goal.

When we move on to the objective of harnessing innovation while meeting our environmental goals, the effect of taxes is even less clear. Do regulators expect to be able to set taxes without any reference to forthcoming technologies which, for example, reduce abatement costs? This would result in higher than necessary pollution reduction as these technologies become available. Regulators would be likely to correct for this by reducing taxes. Unfortunately, any firms tempted by the initially high taxes to develop radically new technologies could be faced with having to write off their efforts when the tax rate falls. Suppose that the regulator does attempt to gather relevant technical and commercial information. The pollution tax is only one of a vast set of costs, and improved technology only one of a vast set of investment opportunities within the industry, all of which the regulator would need to understand in depth. To ensure a constant level of incentive for innovation, the regulator would have constantly to adjust the tax rate as other items such as material costs, turnover, profits and industry structure all fluctuate.

Of course, innovation is not the main concern of regulators considering environmental taxes. If the level of tax can be set with reasonable confidence that its impact on overall emissions will be as expected, it provides a flexible framework within which individual firms can themselves choose the methods and extent of pollution reduction. Unfortunately for technology developers, taxes add a layer of abstraction to the commercial environment within which they operate. They are forced to make judgments about the likely level of demand for new products and processes as a result of the tax. For the reasons set out above, an industry's response to a given tax is likely to be uncertain, making demand for technology correspondingly doubtful and so increasing the risks of innovation. By contrast a soft regulatory target, as used in Japan, sets a concrete technical goal for technology developers.

12.3.2. Tradable Emission Permits

The US scheme to cap SO_2 emissions from electric utilities (described in Chapter 10) provides the first significant test of tradable emission permits (TEPs). It is not too early to attempt to draw some conclusions from early experiences with the scheme.

One of the major advantages claimed for economic instruments is that they require much lower levels of administrative effort than conventional regulations. The SO_2 TEP scheme demonstrates that this advantage may be difficult to secure in practice. In order to establish the legitimacy of the initial distribution of permits, the EPA had to go through an exhaustive assessment of every plant's running costs, tariff rates, fuel type, existing pollution measures and many other factors. This suggests that, to gain industry's willing cooperation, there is no substitute for establishing this kind of informed dialogue – one which also exists in the Japanese soft regulatory system or the Dutch system of voluntary agreements. Once the TEP system is established, this level of dialogue may no longer be necessary, and the administrative effort may indeed be modest. However, any future reduction in the national emissions cap – a process which is analogous to a new round of conventional regulation – would require a repetition of the exhaustive information-gathering exercise. If the EPA has any plans to change the cap, it would be prudent for it to maintain its technical dialogue with the electricity industry.

Another important lesson is that new regulatory mechanisms are not a panacea for poor industry–regulator relationships. The TEP scheme has not produced a clean break from the excessively detailed, restrictive rule-making in which the EPA has always indulged. By opting to phase the scheme in gently, it has reluctantly been forced to adopt its old command and control methods during the transition period.

This TEP scheme also has wider political and institutional consequences. Many state politicians, and officials responsible for regulating the utilities, have failed to grasp the implications of the TEP market. Such schemes play an important role in setting the business environment for the market, and need to state clearly their attitudes to utility investment in pollution control equipment and to retention of trading profits. So far, they have conspicuously failed to do so. TEP schemes elsewhere will need to pay more attention to external agents which can affect the market and the prospects for innovation.

Finally, TEPs and emission taxes bring with them an inherent cost which may be substantial in some cases: TEP schemes will almost certainly require continuous, accurate emission monitoring, because the permit market provides firms with a direct financial incentive to lie about their emissions. Economic instruments have a role – and in all likelihood an increasing role – in environmental policy. However, they should not be regarded as a panacea: care must be taken that they are the appropriate choice and that their implementation does not negate the advantages they apparently offer.

12.4. Voluntary Agreements

'Voluntary agreement' is a term which is being applied to a range of mechanisms which are voluntary in the sense that firms can choose whether or not to take part. The earliest voluntary agreements encountered in this study are the Japanese contracts between firms and metropolitan authorities, dating back to the 1960s. Of greater interest are the systems of agreements covering entire industries. These vary greatly in the extent to which they are 'voluntary'.

Voluntary agreements cannot be seen simply as a further regulatory instrument, to be added to the regulator's armoury. At one extreme, they are indeed being used in this way. At the other, they provide a structured relationship between industry and regulator which sustains a wholesale effort to transform an entire economy. For the most part, this effort is being implemented through the existing system of conventional regulation.

Because of their potential to allow industry to take the lead in determining methods, while allowing policy-makers to set out long-term environmental objectives, certain forms of voluntary agreement are especially well suited to the challenge of sustainable development. In view of their potential importance, this discussion ranges fairly widely over some necessary and desirable features of voluntary agreements, before considering their impacts on innovation.

12.4.1. Background Threats

The Danish voluntary agreement on VOCs illustrates the need for some kind of threat, which may be explicit or implicit, that the industry as a whole will face regulatory action if it does not 'voluntarily' develop a programme to deal with the environmental problem. In Denmark's case, this was an explicit threat of an across-the-board 50% cut in VOC use. The Dutch NEPP forces an industrial sector to offer the highest possible reduction of a pollutant, under the implicit threat that the published, national target for that pollutant might be imposed on that sector, rather than negotiated.

At a more general level, the all-encompassing nature of the NEPP and the broad political support for the goal of a sustainable economy – i.e. the attempt to create policy stability – acts as a powerful implicit threat that the state will always be willing to take action, if industry's own efforts are deemed inadequate. In Denmark, too, there is the implicit threat arising from the state maintaining an effective political distance from the special interests of industry over the last two decades. In Japan, a similar steadfastness on the part of the metropolitan authorities has encouraged many individual firms to make voluntary agreements with these authorities.

12.4.2. Industry-Based Institutions

In the Danish example, in the Dutch NEPP and in the fledgling agreement among electric utilities in the United States, an industry-based institution coordinates each voluntary agreement. It is striking, perhaps, that the prior existence of an appropriate industrial association is by no means essential. In Denmark, the VOC agreement is administered by the national business association, Dansk Industri, within which it is run as a separate project. Within the Dutch NEPP, many of the agreements are run by pre-existing industry associations. However, in the case of the agreement covering the metal manufacturers, there was no appropriate body. An industry association was created especially for the purpose of organizing the agreement, and was supported by government funds.

12.4.3. Coercion of Individual Firms

There may or may not be an explicit threat of regulatory action against an industry, but all the sectoral agreements studied here are voluntary in the sense that an individual firm chooses whether or not to join the agreement. This creates the risk that non-participants will avoid making any efforts to improve their own environmental performance. Inevitably, regulators must act to limit the benefits of being a 'free-rider', by imposing conditions which ensure that there will generally be no advantage in remaining outside an agreement.

In Denmark, licensing authorities can force non-participants to adopt the state of the art in VOC reduction, for their particular industry. Participants in the agreement will in many cases avoid the need to install controls at that stringent level if, for example, their particular industry can make substantial overall reductions through 'good housekeeping' measures. In the Dutch NEPP, participants in the agreements trade off greater or lesser efforts to reduce a range of pollutants. However, non-participants can be forced by licensing authorities to match best practice for *all* pollutants.

In the case of the agreement among US electric utilities to cut CO_2 emissions (administered by the Edison Electric Institute), the EPA has not yet addressed the free-rider problem. However, participants are voicing concerns over the future of the agreement, foreseeing that they will at some point be faced with costly emission reduction measures which non-participants are unwilling to make. Fortunately, the EPA already has a potential regulatory instrument, in the form of a register of emissons and reduction measures, established by the Energy Policy Act. This register could at some point in the future be transformed into a permit trading scheme, or linked to a tax mechanism. The

European experience with voluntary agreements suggests that at some point pressure must be brought to bear on non-participants. The emissions register might provide a novel way of doing this.

12.4.4. Maximizing Information Flows

Voluntary agreements depend for their success on a free flow of information between industry and regulator and, to some extent, between firms within an agreement. Without high quality information on current and projected technologies and markets, regulators would be unable to judge whether the 'offers' coming from industry will achieve the desired environmental objectives. Nor would they be in a position to judge if a particular industrial sector was making sufficient efforts, in comparison to other sectors. A flow of information between participants may be desirable during formulation of industry-level targets, as part of a bargaining process which results in firms making equivalent *efforts*, rather than equal emission reductions.

By emphasizing the process of continuous exchange of technical information between firms and regulators, the system of voluntary agreements set up under the Dutch NEPP reproduces the substance of the flexible regulatory regime created by the Japanese metropolitan authorities. Even the annual 'technical hearings' of the Japanese system are evident in the annual reviews of the 'contracts' which each participating Dutch firm must draw up.

12.4.5. Impacts on Innovation

In most of the systems of voluntary agreement described earlier, an assessment of the potential for innovation is an integral part of setting long- and short-term objectives and monitoring progress. This reflects the origins of the voluntary approach, at least in Europe and the United States, where industry in effect said to the regulators, 'Trust us. Set the direction, and we will explore the best means of getting there.' In response, regulators showed varying degrees of trust, from complete freedom of action in the US electricity utility agreement, to more or less explicit threats of unilateral action if industry fails to make best efforts, in the Netherlands and Denmark.

In the Dutch and Danish system every participating firm is required to set out a detailed plan for reducing emissions. This plan incorporates the likely effects of introducing best practice, improved technologies and speculative technologies, as these are developed. At regular intervals, the firm's actual experience of implementing new techniques, and progress by third party technology developers, is used to adjust future emission targets.

Voluntary agreements appear to be part of a new attitude to the role which industry should play in meeting society's environmental goals. It is an attitude which expects industry to play an active, responsible role, in partnership with government and society. It is an attitude which sits well with the view that our toughest environmental challenges lie ahead of us, and that our solutions must be creative, innovative ones, if they are not to be costly and restrictive.

12.5. Summary

At some point, governments need to choose the regulatory instruments which will, they hope, achieve their environmental goals. If a government is determined to harness innovation, what choices should it make? Regulatory instruments are selected through legislative and regulatory processes. However, the choice of instrument will frequently have less of an influence on the prospects for harnessing innovation, than the political climate of industry–regulator relationships and the nature of day-to-day industry–regulator interactions. These relationships set the context for regulations and deserve more attention from policy-makers and industry. Traditional regulations which specify emission levels or types of equipment are especially sensitive to their context – so much so that regulations which are written identically in two different countries may be radically different in their function and effect. This is a complication which cannot be ignored by policy-makers.

This context-dependency is even more pronounced where government–industry partnership is evoked, especially through a reliance on voluntary agreements. Whether they are known as 'voluntary agreements', 'agreements', or 'covenants', these mechanisms demand a high level of trust between industry and regulator. After all, if they are to be effective, firms will often be required to cooperate in making painful efforts to improve their environmental performance. This cooperation depends on the same conditions which were identified in the previous chapter as the basis for flexible regulatory approaches: good quality dialogue and stability of environmental policy. General guidelines for policy-makers interested in making effective use of voluntary agreements are emerging from experiences with the systems which have already been established.

One general guideline for applying *any* regulatory instrument applies as much today as it ever did: Speak softly and carry a big stick.

Chapter 13

Strategies for Industry–Environment Policy

13.1. Introduction

The previous chapters have addressed the general issue of government of the interface between industry and environment policies, with special consideration of the critical role of industrial innovation. This final chapter will present a tool which will aid strategic management of the interface between national industry and environment policies. I hope that policy-makers who are in the habit of managing policy strategically will find this tool useful, and that those who are not will find it thought-provoking.

Finally, I will attempt to look into the future, put my head above the parapet and recommend an innovation-based strategy for successfully managing an environmentally sustainable economy.

13.2. The Industry–Regulator Matrix

Figure 13.1 presents a matrix of the kind which is often used to analyse business strategies. Here, the matrix represents the possible relationships between industry and regulator[1] on environmental issues. It is composed of the two major factors – political independence of the regulator from industry and quality of dialogue – which differentiate this relationship. (Industry–regulator relationships in each country are analysed in detail in Chapter 11.) As with all diagrams of this nature, its purpose is to provide a *general* insight into the subjects we are interested in, in this case the six countries, and their relationships to one another. It would be wrong to attempt to draw fine distinctions or to attach exact values to the degree of independence or dialogue.

---

[1] For simplicitly, the term 'regulator' as used here includes all those involved in environmental policy-making and regulatory processes in an official capacity, e.g. local or national legislators, senior policy officials and employees of inspectorates.

Figure 13.1 The Industry–Regulator Matrix for Environmental Policy

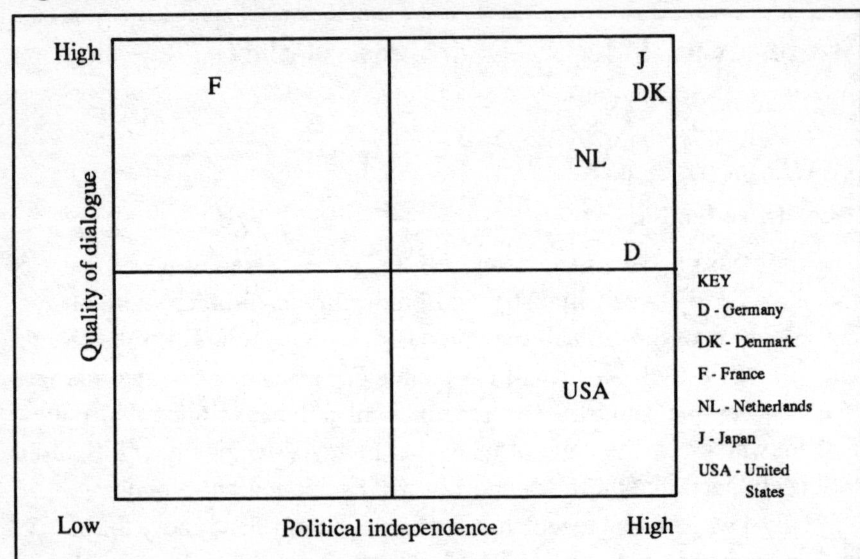

The matrix represents the possible relationships between industry and regulator (all official actors, from environmental legislators to officials of pollution inspectorates) on environmental issues. The *political independence of the regulator from industry* is represented horizontally. The second means of differentiating between countries is by the *quality of dialogue* (the technical competence of regulators, prevalence of personal relationships, industry's willingness to be honest and open in dealings with regulators) during day to day interactions, represented vertically.

Source: Author.

13.2.1. Political Independence of the Regulator from Industry

France, with highly centralized decision-making processes, wide public ownership and a tradition of associating the state with industry, is at one extreme on this measure. Japan, with its locally driven environmental policy, and Denmark, where the Environment Ministry is extremely powerful and great efforts are made to involve all sectors of society in policy formulation, are at the other extreme.

In Germany and the Netherlands the political relationship is currently shifting. I will show how the matrix can provide insights into these shifting relationships a little later in this chapter. For the moment, Figure 13.1 shows the situation as it was in the late 1980s, when Germany was pushed to one extreme by regulators who would on occasion refuse to have any dealings with industry, and by the presence in the legislature of one of the world's largest 'Green' political parties. In the Netherlands political concern for the environment was high but this was counterbalanced by the political influence of industries which felt they had made great improvements, at great cost, since the 1970s. In the United States, the enthusiasm of Congress for environmental legislation has been tempered only marginally by industry's massive lobbying efforts.

13.2.2. Quality of the Industry–Regulator Dialogue

The existence of a flow of information between industry and regulator does not, in itself, necessarily constitute an effective dialogue for the purpose of harnessing innovation. The United States presents examples of consultation between the two taking place in an atmosphere of distrust, making the information of little use. In general, the industry–regulator dialogue in the United States is poor, reducing the tendency for innovative responses and increasing the costs of environmental regulation. In Germany, too, the dialogue is relatively poor, despite a good state–industry dialogue in many other policy areas.

At the other extreme, Japanese industry benefits from close, continuous, well-informed dialogue with local authority officials. So, too does Denmark, where Environment Ministry officials have cultivated a climate of close cooperation with industry, over the last two decades.

In France, too, an informed relationship exists, but in this case it is important to draw a careful distinction between the regulator and other parts of the government. The French Industry Ministry has a very close and well-informed relationship with industry. Indeed, in many cases they are effectively one and the same. Once a political decision has been made to pursue a given environmental objective, this close relationship is put to good use, for example through the practice of applying the site licensing system through regional offices of the Industry Ministry. In this way, details of environmental regulations and their implementation and enforcement are sensitive to the workings of industry, and innovation is allowed to play a role. However, there are indications that the knowledge which the Industry Ministry holds is not brought to bear in the earliest discussions of environmental issues, when tensions exist between the Industry and Environment Ministries. The outcome is a relative complacency on environmental issues, but an ability to exploit innovation when regulatory action becomes necessary.

13.3. Emergent Strategies for Industry–Environment Policies

One of the main features of a matrix analysis is that it allows us to identify various strategies for dealing with the issue at hand. Looking again at the industry–regulator matrix, a set of possible strategies for governments to pursue, when dealing with the interface between industry and environment policies, begins to emerge. These strategies are represented by the four quadrants of the matrix, and are set out in Figure 13.2.

I have summarized the extremes of low or high political independence from industry on environmental issues with the terms (environmentally) 'complacent' and (environmentally) 'responsible'. Readers would be correct in detecting a certain bias towards environmental concerns, rather than industry's concerns, in this formulation. Similarly, according to the quality of dialogue between its regulatory institutions and industry, a country will adopt either an 'innovator' or a 'non-innovator' approach to meeting environmental objectives.

13.3.1. Industry–Environment Strategies

For the policy-maker, these strategies provide a range of options for tailoring industry and environment policies to national political, economic and cultural conditions and for managing the most pressing problems associated with environmental regulation.

- *Complacent Innovator*

This strategy can be a successful one for those countries with a relatively low level of environmental concern. When environmental issues do have to be tackled, the methods for doing so are likely to be in keeping with industry's capacity to respond technologically and organizationally. This will tend to

Figure 13.2 Strategies for Industry–Environment Policy

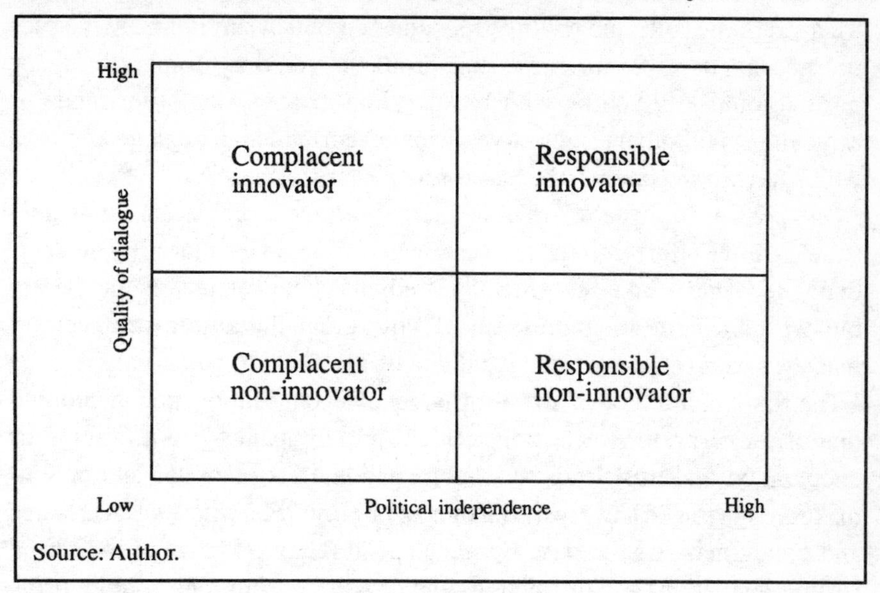

Source: Author.

favour lowest-cost, innovative responses to environmental challenges and so avoid any loss of competitiveness for the industries involved.

With this strategy the greatest disadvantage is that of political risk. Environmental issues may suddenly assume increased political importance, exposing the close industry–state relationship to criticism and perhaps hampering socially optimal decisions. The political risk may arise domestically, or, increasingly, internationally. At first sight it seems possible for a country in this situation to be able to make a rapid transition to the less risky 'responsible innovator' strategy. In practice, the institutional strengths of the industry–regulator dialogue are likely to be inseparable from the 'cosy' political relationship between the two.

This may be the case in France. A greater political separation would probably have little impact on the *formal* structure of the regulatory system. However, initially at least, one would expect industry to be less cooperative with an Environment Ministry which has acquired greater political clout than it has been with the Industry Ministry. This could disrupt the effective dialogue which now exists.

• *Responsible Innovator*

This strategy avoids the political risk inherent in new environmental issues or increased environmental consciousness, either domestically or internationally. It shares with the previous strategy the advantages of achieving environmental objectives at lowest cost, through flexible processes which allow innovation to be harnessed.

One possible drawback, which it shares with the 'complacent innovator' strategy, is the effort required to maintain the quality of the industry–regulator dialogue.[2] This is an issue which the study has not examined in any depth, but we can form an impression of how great the administrative and management costs of an effective dialogue are likely to be.

The first point to note is that the dialogue depends more upon an attitude than a set of prescribed actions which must be paid for. An attitude of cooperation and trust leads to local or national inspectorates and policy-makers engaging in a two-way dialogue with firms. Elsewhere, inspectorates and policy-makers still exist, but they spend relatively more of their time issuing regulations and instructions and devising schemes for forcing firms to comply with them.

Even if there is some additional cost to engaging in this dialogue, as seems likely, common sense suggests that it is outweighed by the benefits. Poor dialogue can lead to a 'mistake' such as the premature creation of an electric vehicle industry and the diversion of huge sums of public money to related R&D and infrastructure programmes. Even if hundreds of additional officials were needed to improve communication with industry, the cost is likely to be small by comparison with the opportunity cost of bad policies.

• *Responsible Non-Innovator*

This is a strategy with little to commend it. Countries end up here by accident rather than design, and then agonize about how to escape. The answer is, with great difficulty.

This is the area where regulatory approaches have been rule-bound and costly for industry. Where the affected industries compete in international

---

[2] However, the good dialogue of the Complacent–Innovator strategy will normally exist for other reasons: environmental issues merely 'borrow' these channels of communication. Unusually, Japan's 'technical hearings' dialogue mechanism is created solely for the purpose of dealing with environmental issues, so their 'cost' must be attributed to environmental policy.

markets, worries regarding international competitiveness add to political anxieties over the burden of regulations. The recessions which are part of the normal business cycle will bring calls for deregulation and less concern for the environment to periodic crescendos. This creates a dichotomy between environmental protection and cost, producing policy swings with each change of political leadership.

Escape from this costly, politically charged strategy is possible by becoming complacent or innovative, or some combination of the two. For developed countries, trading off economic costs against environmental protection does not seem to be realistic. Only in the United States, during Reagan's first presidential campaign and subsequent term of office, has there been a serious attempt to persuade the electorate of the need for a transition from the 'responsible' to the 'complacent' strategy. In the event, the only practical outcome was a series of delays in new legislation, not a wholesale rolling back of environmental policy.

The only practical option is to become more flexible and innovative, by creating both the policy stability and the institutions which lead to an effective dialogue. This is inevitably a slow, laborious process and for that reason alone it is a difficult one. The prospects for the United States achieving a successful transition to 'responsible innovator' – a process which it has recently begun – will be discussed shortly.

• *Complacent Non-Innovator*

This is the realm of the less developed countries (LDCs), and for them it can be a successful and sensible strategy, although it brings with it a continuously increasing risk.

In many LDCs, economic liberalization is causing rapid economic growth. Even if they wanted to, LDC governments are often unable effectively to regulate labour conditions, health and safety or environmental performance. Redressing such weaknesses in institutional capacity is an objective of international aid agencies, such as the UNDP. At the same time, LDC governments are generally far more concerned with promoting industrialization than with protecting the environment. They devote much of their resources to assisting their industries to find foreign partners or acquire foreign aid. Politically, they tend to be closely associated with their industries. As long as advances in living standards outstrip any environmental damage, and the population approves of this approach, the strategy can rightly be judged a success (international environmental impacts aside).

However, there are many examples of LDC governments turning a blind eye to their people's concerns about the environment. As LDC countries grow richer these concerns are likely to grow, to the point where action is demanded. To satisfy a sudden demand for action on a broad range of environmental issues, a government would almost certainly have to sever its close political relationship with industry. While this can be done almost overnight, it takes much longer to create the conditions for an effective dialogue with industry. Concrete action on the environment is therefore likely to include inflexible and costly regulatory impositions on industry.

Many LDCs are, therefore, running the risk of being forced to make a rapid transition to the 'responsible non-innovator' strategy, and suffering the consequential costs to industry. There are reasons for believing that this may be more rapid and more painful for LDCs than it was for the United States, for example. First, the US regulatory model exists in its entirety, ready to be adopted by any nation with a new-found concern for the environment. By contrast, the full set of legislative and administrative mechanisms which now exist evolved over decades in the United States. Secondly, several of the countries in this study, including the United States and Japan, are taking active steps to export their regulatory regimes to developing countries, increasing the likelihood that the industry–environment strategies of the target LDCs will indeed shift significantly.

Figure 13.3 shows two contrasting paths for LDC strategy transitions. Path 1 is the (potentially rapid) shift brought about by adopting a US-style regulatory regime. Path 2 is potentially a more desirable route to the 'responsible innovator' strategy, but here, too, there are pitfalls. It is certainly important to create an effective industry–regulator dialogue. This is where aid bodies can fruitfully concentrate their efforts in institutional capacity-building. Ideally, the LDC government would gradually set out a clear environmental agenda which creates an effective political distance from industry. Environmental objectives would be set after flexible mechanisms which allow innovation to be harnessed have been established. The danger lies in the possibility that the government's dialogue with, and understanding of, industry will simply cement close political relationships, ruling out effective distancing on environmental issues.

Figure 13.3 Strategy Transitions

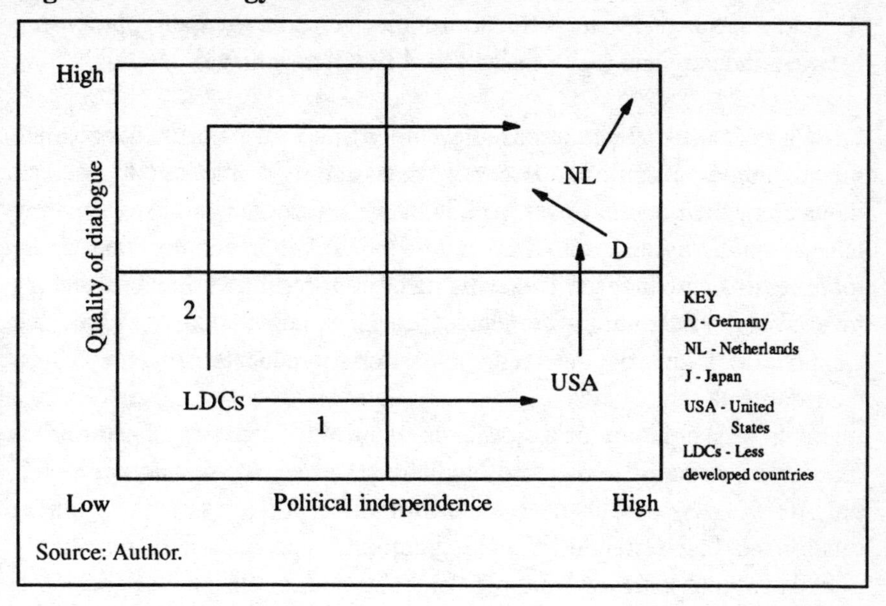

Source: Author.

13.3.2. Understanding Recent Shifts in Industry–Environment Policy

Of the countries in this study, Germany, the Netherlands and the United States have been experiencing significant shifts in industry and environment policies in recent years (see Figure 13.3).

Germany In Germany, industry's complaints about the costs of environmental regulations reached a peak in the recession of the early 1990s. Politically, industry and the state drew closer together, resulting in an informal moratorium on new regulations, causing great concern in the Environment Ministry. Fortunately, there are also signs of contrition from the ministry over its past unwillingness to consult with industry. This has produced a conscious effort by the ministry to improve its dialogue with industry, for example, through open discussions with the federation of small businesses.

The industry-led waste recycling agreement, and proposals for similar agreements on product recycling, are an indication that a more innovative regulatory climate may develop in Germany. On the other hand, confrontational statements by the former Environment Minister, Klaus Töpfer, and vehicle industry representatives, on the subject of vehicle recycling regulations, suggest that a shift to a more cooperative, open relationship between industry and regulator will be difficult to achieve.

The United States Regulators in the United States have begun to recognize the procedural origins of the excessive costs and rigid practices imposed on industry by their regulatory system. During the Clinton administration they have adopted a number of initiatives with the explicit intention of consulting more closely with industry and avoiding or improving traditional regulatory mechanisms. These initiatives include 'regulatory negotiation' processes (see Chapter 8), voluntary agreements and efforts to educate firms about best environmental practice.

This is the beginning of a slow process of transformation of the role of industry in standard setting and regulation. During this transformation (if the effort can be sustained), there will be twin pressures on the political relationship between industry and regulator. On one side, industry is likely to call for more leniency from regulators, as a trade-off for their increased cooperation. On the other side, environmental lobby groups will interpret a softer, more consultative regulatory approach as a weakening of the independence of the regulator from industry. If any such weakening *does* occur, their protests will be so much the louder.

The Netherlands Since the late 1980s, the Netherlands has shifted in the opposite direction to Germany. This has been a direct consequence of the government's commitment to a long-term plan for creating a sustainable economy, and the means it has chosen for achieving this. These two factors deserve special attention.

13.3.3. The Strategic Effect of Sustainable Economy Plans

Chapter 4 describes the origin of a plan to make the Dutch economy environmentally sustainable and outlines the comprehensive policy for achieving this goal, the National Environmental Policy Plan (NEPP). By stating long-term, concrete goals for environmental policy (up to 2010, initially) and by basing them on the objective of environmental sustainability, the government has created a policy framework which is stable, gains credibility as time passes and increases and formalizes political independence from industry.

It is possible to interpret the effect which a stable environmental policy has on the behaviour of firms as just one example of a more general feature of policy-making. It may be that policy objectives can be achieved more easily – at lower cost, with less effort to force compliance, and so on – whenever they are grounded in a consistent and simple framework. One area where this phenomenon is apparent is monetary policy: in Germany an independent central bank with a simple objective of suppressing inflation gives investors the reassurance to offer long-term Deutschmark loans at low rates of interest. In many other countries, monetary policy is influenced by short-term political considerations. Long-run inflation rates, and therefore the cost of borrowing long term, are higher. The benefits of low borrowing costs to the German economy have been immense but the remarkable fact is that these benefits are free, depending only on the sentiment of investors.

The Industry–Regulator matrix gives an insight into the impact of the Dutch NEPP on the behaviour of industry and others. Once it has gained wide political and public acceptance, a sustainable economy plan limits the strategic space which a nation can inhabit, as indicated by shaded area 'A' in Figure 13.4. This period of gaining acceptance and political credibility can be likened to the period following the announcement of some new monetary target during which financial markets reserve their judgment. In many cases, such as the various measures of money supply pursued in the United Kingdom during the 1980s, these policies are abandoned by politicians when the going gets tough. Germany's steadfastness is very much an exception.

It would not be surprising if those affected by the Dutch NEPP were to take a similar approach and withhold their full cooperation until they were convinced of the seriousness of the goal of creating a sustainable economy. It is to be hoped that Dutch politicians do not squander the political capital they have invested in doing so before it bears fruit. The practical demands of the NEPP on Dutch industry are unprecedented. Attempting to meet them without industry's full cooperation, indeed without industry playing a leading role, would be a recipe for economic disaster. Given time the NEPP should support the necessary sentiment for this cooperation to take place.

Figure 13.4 Strategic Effects of National Plans (A) and Voluntary Agreements (B)

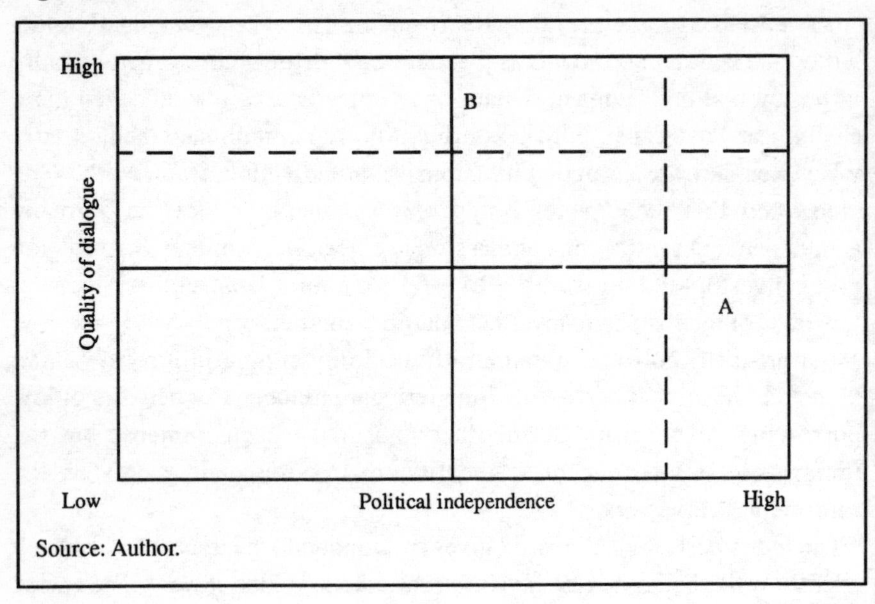

Source: Author.

13.3.4. The Strategic Effect of Voluntary Agreements

As area 'A' in Figure 13.4 indicates, having a plan for a sustainable economy leaves open the option to pursue it through imposing regulations without regard to the circumstances of particular industries or firms, and therefore with little chance of harnessing innovation. Clearly, when environmental policy is as far-reaching and ambitious as the NEPP, it is more important than ever to achieve a maximum dialogue and level of trust between industry and regulator, i.e. to choose the 'innovator' rather than the 'non-innovator' strategy.

The Dutch system of voluntary agreements should do precisely this. The agreements formalize industry–regulator dialogue at all stages of the regulatory process. Prospects for technical innovation over a range of time scales are used to set formal, but flexible, targets for environmental performance. The effect of voluntary agreements of this type is to confine industry–environment policies to the strategic space indicated in Area 'B' in Figure 13.4.

Taken in combination, the system of voluntary agreements and the sustainable economy plan fix a nation firmly within the 'responsible innovator' strategy. An additional advantage is that these two measures are mutually reinforcing: Chapter 11 and Chapter 4 explain the role of the system of voluntary agreements in sustaining industry's support for the NEPP and the importance of the NEPP in persuading industry to respect the processes embodied in the system of voluntary agreements.

13.4. Innovation and Sustainable Development

At the Rio Earth Summit in 1992, most countries committed themselves to the principle of 'sustainable development', but agreed to differ on exactly what that means. Many hoped to reach agreement on mechanisms which would arrest global environmental degradation, but this was not possible. The measures which were agreed are only tentative first steps towards environmental sustainability.

Let us imagine that a few years from now those same countries feel a greater urgency to move towards environmentally sustainable development, perhaps because a rapid global warming effect has been confirmed. In what ways will our modern industrialized economies need to be transformed to achieve this, and how can this be done in a manner which minimizes the impact on our standards of living? Politicians and policy-makers are ultimately responsible for articulating and leading this effort. Businesses can choose to cooperate (and show leadership themselves) or to obstruct. This challenge is simply the familiar one of environmental protection, writ large. Understanding the political, social and organizational processes which provide most gain, for least pain – through innovation – is now more important than ever before.

13.4.1. Industry's Role: Proactive and Responsible

In recent years, many of the world's largest industrial firms have undergone a transformation in their public attitude to the environment. For some, this has been accompanied by a general transformation in their internal culture. Multinationals such as 3M and Dupont have found that by anticipating sources of pollution and eliminating them at the design stage, they can save on raw materials, waste remediation equipment and pollution charges. This reduces their capital requirements, increases their efficiency and saves money. In the run-up to the Earth Summit, and subsequently, organizations representing major firms, such as the Business Council for Sustainable Development,[3] have argued that they have now developed a sense of responsibility for the environment and can be left to pursue maximum environmental performance, free from interference by public authorities. They point to the new management and production concepts they are exploring, such as 'design for the environment' and 'industrial ecology', as proof of their commitment and sense of responsibility.

It would be unwise blindly to trust industry's statements about its new attitude to the environment, but current business trends towards increased worker and stakeholder participation are consistent with these claims. Governments and regulators should support these trends through appropriate policies. Stable long-term policies and effective dialogue can create a culture which rewards proactive, responsible firms and reduces the risks of investing in innovative solutions.

13.4.2. The Role of Governments: Biting the Bullet

If sustainable development is to have any meaning, national governments must face up to the issue of how to make their economies environmentally sustainable. Of the countries in this study, the Netherlands has taken this process furthest. First it determined its environment's carrying capacity for waste and pollutants. This analysis forced the country to conclude that a radical transformation of the economy would be essential to reduce pollutant and waste loads to sustainable levels.

Because the Dutch authorities decided not to wait for other countries to make similar efforts to create a sustainable economy, they recognized the potential risk of imposing massive costs on their industries which would harm their international competitiveness. By demonstrating firm political commitment to a sustainable economy and involving industry intimately at every stage, they have secured industry's cooperation in a process of transformation through innovation. They believe this to be the least-cost approach.

---

[3] Now merged with the International Chamber of Commerce's World Industry Council on the Environment to create the World Business Council on Sustainable Development.

Elsewhere, some countries are taking tentative steps to follow the Netherlands, but have so far been held back from full-blooded commitment through fear of the costs of being members of a very small club. Japan recently enacted its 'basic plan for the environment', which has the potential to evolve into a national plan for a sustainable economy. Germany plans to work towards a sustainable economy through product recycling requirements, but this policy may founder through a failure adequately to involve industry. Denmark has begun an inventory of how all materials flow through society, as a prelude to possible national targets for consumption. However, as a small export-based nation, Denmark feels particularly vulnerable to the risk of competitive disadvantage.

It is open to other countries to continue paying lip service to sustainable development, while ignoring its practical implications. Fortunately, the political processes generated by the Earth Summit provide a regular forum for those countries which take unilateral action to parade their successes and so pressurize the others. Governments which do this have the freedom to formulate policies and processes which encourage flexible and innovative responses from industry. Those who wait too long may be overtaken by international pressure for action, risking rapid, disruptive policy changes and losing the opportunity gradually to develop the dialogue mechanisms which can limit the pain of the transition to sustainability.

Sustainable development should be regarded as a positive force. It has the potential to be the focus of a coherent, comprehensive agenda for the environment, as in the Netherlands. It is inherently long-term and all-encompassing, and so lends itself to establishing a social consensus, in a way which individual environmental problems cannot do because they can be 'solved' and so disappear from popular consciousness. Taken seriously, sustainable development provides an escape route from the reactive, crisis-driven policies of the past to a stable policy framework where the role of industry is clear. As this clarity and stability emerge, risk recedes, constructive dialogues are established and innovation is not far behind.

Index